建筑立场系列丛书 No.26

C3

锚固与飞翔——挑出的住居
Podia, Plinths and Flying House

中文版

韩国C3出版公社 | 编

于风军 王平 王凤霞 高翔 郑海荣 阿斯亚·阿不力米提 | 译

大连理工大学出版社

4 资讯

- 004 亲水设施_Bureau FaceB
- 006 凤凰城瞭望塔_BIG
- 010 博科尼大学的新城市校园_SANAA
- 012 郊野游乐园_Openfabric + dmau
- 014 中世纪波斯尼亚历史公园_Filter Architecture
- 018 镜子实验室2.1_VAV Architects
- 020 格兰富大学学生宿舍_CEBRA

22 墙体工艺

- 022 修复过去的废弃建筑：对混合结构的颂扬_Nelson Mota
- 026 柯尔顿公寓_3ndy Studio
- 038 塔利亚剧院_Gonçalo Byrne Arquitectos + Barbas Lopes Arquitectos
- 046 Potxonea住宅_OS3 Arkitektura

52 锚固与飞翔——挑出的住居

建筑立场系列丛书 No.26

- 052 锚固与飞翔——挑出的住居_Silvio Carta
- 058 坦格尔伍德2号住宅_Schwartz/Silver Architects
- 066 Algarrobos住宅_Daniel Moreno Flores + José María Sáez
- 076 BF住宅_OAB – Office of Architecture in Barcelona + ADI Arquitectura
- 086 X住宅_Cadaval & Solà-Morales
- 096 贝兰达住宅_Schmidt Arquitectos
- 106 素风宅_acaa/Kazuhiko Kishimoto
- 116 Hanare住宅_Schemata Architects
- 124 纳克索斯岛避暑别墅_Ioannis Baltogiannis + Phoebe Giannisi + Zissis Kotionis + Katerina Kritou + Nikolaos Platsas

132 Héctor Fernández Elorza建筑师事务所

- 132 假若我们相见_Héctor Fernández Elorza
- 134 材质的高密度_Jesús Donaire + Héctor Fernández Elorza
- 138 贝内西亚公园
- 150 双子广场
- 160 Valdefierro公园
- 172 细胞和遗传生物学学院

- 184 建筑师索引

News

004 Water at-traction _ Bureau FaceB
006 Phoenix Observation Tower _ BIG
010 A New Urban Campus for Bocconi University _ SANAA
012 Into the Wild _ Openfabric + dmau
014 Historical Park of Medieval Bosnia _ Filter Architecture
018 Mirror Lab 2.1 _ VAV Architects
020 Grundfos College Student Dormitory _ CEBRA

Wall Graft

022 *Revamping the Derelicts of the Past: In Praise of the Hybrid _ Nelson Mota*
026 Corten Apartments _ 3ndy Studio
038 Thalia Theater _ Gonçalo Byrne Arquitectos + Barbas Lopes Arquitectos
046 Casa Potxonea _ OS3 Arkitektura

Podia, Plinths and Flying House

052 *Podia, Plinths and Flying House _ Silvio Carta*
058 Tanglewood House 2 _ Schwartz/Silver Architects
066 Algarrobos House _ Daniel Moreno Flores + José María Sáez
076 BF House _ OAB – Office of Architecture in Barcelona + ADI Arquitectura
086 X House _ Cadaval & Solà-Morales
096 House in Beranda _ Schmidt Arquitectos
106 Wind-dyed House _ acaa/Kazuhiko Kishimoto
116 Hanare House _ Schemata Architects
124 Summer House in Naxos _ Ioannis Baltogiannis + Phoebe Giannisi + Zissis Kotionis + Katerina Kritou + Nikolaos Platsas

Héctor Fernández Elorza

132 *Incase We Meet _ Héctor Fernández Elorza*
134 *Intense Material Density _ Jesús Donaire + Héctor Fernández Elorza*
138 Venecia Park
150 Twin Squares
160 Valdefierro Park
172 Faculty of Cellular and Genetic Biology

184 Index

4 城市便利设施　URBAN AMENITY

亲水设施 _Bureau FaceB

新建筑新用途：人行桥上水上漫步

该项目位于市中心，是巴黎市一处重要的景点。然而，项目设计内敛、不张扬，让人时而感觉水上泛舟，时而感觉桥上漫步。相反，塞纳河被人们视作一处超越时间的地方，向人们讲述着故事与历史，并将时间与空间联系起来：水景迷人。

这座新桥如神来之笔在河面上轻轻一挥，同河水嬉戏玩耍。桥的设计没有应用传统的压缩工艺，而是应用牵引力潜能这一新的设计。钢缆固定在河两岸的起拱面上，交织成网，将混凝土块如珠子一样串在一起。

整个设计流畅，用途新颖多样。到达河对岸有两种方式。

一种看起来有些"危险"，窄窄的桥面犹如喜马拉雅山脉上面的人行天桥；另一种方式是通过一个开阔的空间，人们可以休闲漫步，在接近水面的地方坐下，安静地休息，吃午餐，近距离地欣赏塞纳河风光，以一种别样的视角看巴黎。

Water at-traction

New Structure for New Uses : a Pedestrian Bridge to Stroll along the Water

It is in the heart of the city, as one of its major attraction. However, you can barely feel it, maybe on a boat, a little bit on bridges. On the contrary the river Seine has to be seen as an out-of-time place, telling you the stories and history, a link through time and space: the water attraction.

This new bridge has to be seen as a light stroke, a thin roadway flirting with the water. Instead of using traditional technics based on compression, it uses a new design, using the potential of traction. Steel cables, strung between the banks by springs, generate a mesh on which concrete beads are threaded.

This fluent area enables new uses. The crossing can be done in two ways. Through a "perilous" one: the very narrow deck gives the feeling of a Himalayan footbridge. Through a space for strolling: the generous space near the water allows to sit, to rest quietly, having lunch, enjoying the proximity of the river and offering a unique perspective on Paris.

项目名称：Water at-traction / 设计时间：2012
地点：Paris, France
建筑师：Camille Mourier, Germain Pluvinage
项目团队：Arnaud Malras, François Marcuz, Camille Mourier, Germain Pluvinage

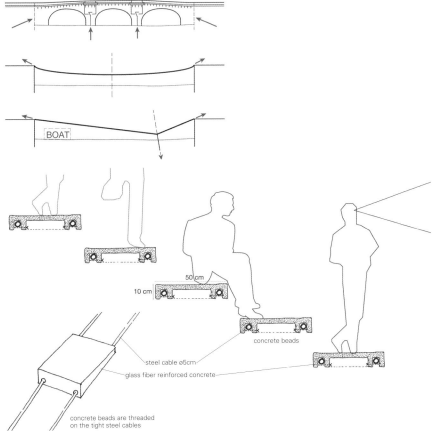

凤凰城瞭望塔 _BIG

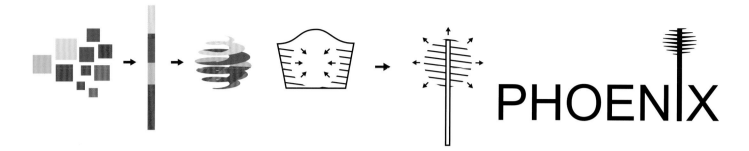

Novawest（一家精品房地产企业）委托BIG建筑师事务所在美国亚利桑那州凤凰城设计一座城市新地标——123m高的多功能瞭望塔。

瞭望塔位于凤凰城市中心，为城市的天际线又增加了一座靓丽的建筑，人们可以在上面鸟瞰整座城市的壮观景色及其周围绵延的山脉和绚丽的落日，该塔以其独特的设计成为游客和市民的必游之地。

正如Novawest公司的布莱恩·斯托所说，对凤凰城市中心来说，该项目的建造可谓是天时地利，Novawest公司知道凤凰城就需要这样非凡的建筑。BIG建筑师事务所的设计卓尔不群，集形式与功能于一体，从此，凤凰城的天际线将大不一样，登顶之上将给游客带来千载难逢的经历与体验。

未来的瞭望塔由钢筋混凝土打造而成，高耸入云，开放的螺旋状球体位于顶部，像一颗大头针插在地图上标明一个位置所在。螺旋状的球体空间内设有展厅、零售商店和娱乐空间，游客可以乘坐三部玻璃观光电梯上上下下，游览瞭望塔，并一览凤凰城全景。

游客从瞭望塔顶部沿着螺旋状人行步道向下，可以连续不断地体验建筑中的所有设施，360°动态观赏凤凰城和亚利桑那州风光。

根据BIG建筑师事务所的建筑师Bjarke Ingels所说，正如季风、哈布沙尘暴和周围亚利桑那州的群山一样，被称为"大头针"的瞭望塔成为一个参照点以及一种途径，游客在上面闲庭漫步就可以让目之所及成为移动的风景。就像纽约的古根海姆博物馆，游客沿着中央开放空间缓缓而下，就有一种独特的艺术体验，而游客在"大头针"上的游览是由内而外的，可以眼望凤凰城整座城市和周围风光而充满无限遐想。"大头针"就像一个天体一样漂浮在城市上空，使游客可以一层层盘旋而下，仿佛悬浮在半空中，尽享动态三维立体体验。

螺旋状的布局将不同的功能区域和交通流线结合成为一个连续、动态的旋转空间，其面积大小是根据游客的行走轨迹而设定的，为游客带来一种独特的眺望四周景色的体验。螺旋状人行步道宽窄不一，入口处最窄，中间处最宽，然后渐窄，出口处又变回最窄。

球体内部各功能区之间的划分不是通过厚重的墙体，而是通过坡度和高度的舒缓改变，从而保持了整个空间的连续性，使用来举办展览和活动的空间灵活多变。

当游客游览到球体中部时，可以选择乘坐电梯返回地面结束本次游览，也可以选择继续到位于球体低层的餐厅层就餐。这一行程就像是穿过行星中心和从北极到达南极的旅程。

瞭望塔底部是公共广场，游客可以在此纳凉休息，有水景，有少量零售商店，还有一块地下登塔排队区。瞭望塔的设计将太阳能技术和其他技术综合在一起，将成为可持续利用能源的设计典范。

Phoenix Observation Tower

BIG was commissioned by Novawest, a boutique real estate enterprise to design a 123m tall mixed-use observation tower to serve as symbol for the city of Phoenix, Arizona.

Located in downtown Phoenix, the observation tower shall add a significant structure to the skyline from which to enjoy the city's spectacular views of the surrounding mountain ranges and dramatic sunsets. The tower can be a destination event providing the unique features to tourists and citizens.

According to Brian Stowell from Novawest, this is the right place and the right time for a signature project for downtown Phoenix and they knew the design needed to be something extraordinary. BIG has delivered something exceptional, blending form and function in a way that will change the local skyline forever and will give visitors a once-in-a-lifetime experience.

The future observation tower is conceived as tall core of reinforced concrete with an open-air spiral sphere at its top, resembling a pin firmly marking a location on a map. The spiraling sphere contains exhibition, retail and recreational spaces which are accessed via three glass elevators connecting the base with the summit and offer panoramic view of the city and the tower's programs as visitors sascend or descend. Walking downwards from the top through a spiral promenade, the visitors experience all of the building's programs in a constant motion, while enjoying dynamic 360 degree views of the city of Phoenix and the Arizonian landscape.

According to Bjarke Ingels, the architect of BIG, like the monsoons, the haboobs and the mountains of the surrounding Arizonian landscape, the Pin becomes a point of reference and a mechanism to set the landscape in motion through the movement of the spectator. Like the Guggenheim Museum of New York offers visitors a unique art experience descending around its central void,

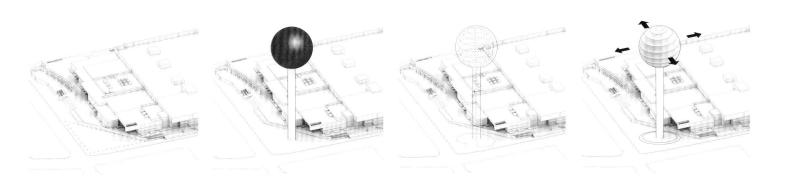

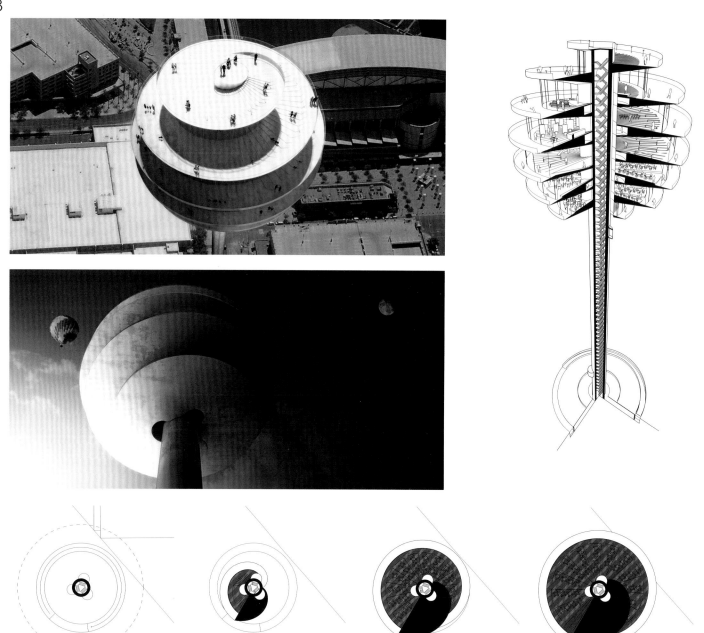

| 一层 first floor | 二层 second floor | 三层 third floor | 四层 fourth floor |

the motion at the Pin is turned inside-out allowing visitors to contemplate the surrounding city and landscape of Phoenix. Like a heavenly body hovering above the city, the Pin will allow visitors to descend from pole to pole in a dynamic three dimensional experience seemingly suspended in midair.

The spiral layout combines the different programmatic elements and the circulation into a continuous dynamic twirling space which is proportioned according to the movement of the visitors, producing a unique viewing experience of the surroundings. Instead of a constant width, the spiraling promenade starts from zero at the point of arrival, reaches its maximum width at the middle, and shrinks back to zero at the point of departure.

Separation between the programmatic elements within the sphere happens not through physical barrier-walls, but softly through the slope and the height difference to preserve a total continuity and create a flexible space for exhibitions and events.

Once the visitors reach the middle of the sphere, they can choose to either conclude their journey by taking the elevator back to the ground, or continue to the restaurant levels at the lower hemisphere. The motion resembles a journey through the center of a planet, and a travel from the North to the South Pole.

The base of the tower will serve as a public plaza offering shade, water features and a small amount of retail together with a subterranean queuing area. The tower will serve as a working model of sustainable energy practices, incorporating a blend of solar and other technologies.

项目名称：Phoenix Observation Tower
地点：Phoenix, Arizona, USA
建筑师：BIG
项目团队：Thomas Fagan, Aaron Hales, Ola Hariri, Dennis Harvey, Beat Schenk
项目指导：Iannis Kandyliaris
合作者：MKA, Atelier10, Gensler, TenEyck
甲方：Novawest
总建筑面积：6,500m²
设计时间：2012

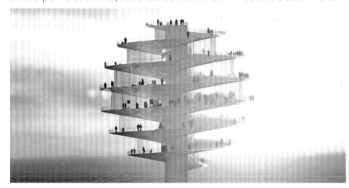

五层 fifth floor

六层 sixth floor

七层 seventh floor

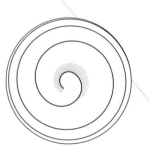
屋顶 roof

博科尼大学的新城市校园_SANAA

米兰博科尼大学新校区以前是个奶制品厂。博科尼大学为新校区的建造举行了一场国际设计竞赛,一些在国际上享有盛名的建筑事务所应邀参加了竞赛。同时,博科尼大学也为学习建筑和建筑工程专业的学生们举行了设计竞赛。以《建筑电讯》创始人彼特·库克先生为首的评审团比较了所有参赛作品之后,最终评定SANAA的设计作品胜出。整个项目占地35 000m²,预计于2018年建造完成。

透明、亲近自然以及愿景

博科尼大学的新城市校园项目解决了具有挑战性的城市环境问题,既营造了卓越的学术环境又使用了极其节能环保的建筑体系。南面是车水马龙的大街,西面是一系列面向大道的中等规模的住宅建筑,东面是公园,北面是一些高层住宅楼和大学建筑混杂在一起,与现有的博科尼大学校园相连。为了把如此多样的城市形态和谐地融为一体,建筑师们把不同功能区域纳入相对独立的单元格中,并对其在基地中的位置进行仔细思考,从而使整个项目犹如一座小型城市。

新的SDA博科尼管理学院入口位于整个项目基地的北面。入口建筑设有一个两层高的透明大厅,穿过入口建筑就可以到达这个综合设施的内部庭院系统。南面,娱乐中心成为校园与车水马龙的大街之间的缓冲地带。建筑师将学生宿舍修建在项目基地中较为安静的东面。西面建有研究生楼和行政大楼,内有普通教室、礼堂教室和研讨室。学生从教室出来就可以马上到研讨室一起讨论课程内容。礼堂教室内的桌椅皆可移动,教授可以在学生中间走动。

弧形的建筑平面与教室的形状保持一致,贴切而自然。为了使建筑内的路径便利通畅,建筑师将每个单元格连接起来,这样就在学校与外界之间形成了一个连续不断的边界。所有建筑都建于通透的门廊之上,支柱、透明房间和树木的序列非常清晰。封闭的花园、庭院和门廊是米兰这座城市显著的建筑特点,为人们在室外社交、学习、聚会、会谈提供了安静祥和的场所。

每个单元格的空间都设计得比较狭窄,室内完全沐浴在自然光之中,并设有通向公园庭院的大型洞口,达到最优化的自然通风,同时降低制冷和照明负荷。外墙50%不透明,另外50%为透明玻璃,因而取得了最理想的保温效果并降低了成本。在水资源利用方面,建筑师采用了地下水或循环雨水的能源策略。

该项目旨在设计一个学生、教师和来访者都能成为蒸蒸日上的学术社区一分子的大学校园,感受其无限的透明感、与大自然的亲近感以及美好愿景。建筑师们希望该项目的灵活性和连通性能使博科尼大学带着雄心抱负一路畅通无阻地发展。

A New Urban Campus for Bocconi University

Bocconi University in Milan held an international competition for ideas for construction of a new campus on the site of the city's former dairy. The competition formula, a competition on invitation, compared the ideas of a number of the world's most prestigious architectural studios and also included a competition for students of architecture and construction engineering. The panel of judges, under *Archigram* founder Sir Peter Cook, chose SANAA as the winner. The project will be built by 2018 on a 35,000m² plot.

Transparency, Closeness to Nature, and Vision

The project for a new urban campus for Bocconi University solves a challenging urban condition while fostering an excellent academic environment and using an extremely energy efficient building system. On the south there is a highly trafficked thoroughfare, on the west a series of medium-scale residential buildings facing the large road, on the east the park, and on the north a mixture of tall residential and university buildings linked to the existing Bocconi Campus. In order to harmoniously engage such urban diversity, the architects have organized the program in separate cells and sensitively located them on the site, for the project to acquire an urban scale.

The entrance to the new SDA Bocconi School of Management stands on the north side of the site. The entrance building has a double-height transparent lobby. Through the entrance building, it is possible to access the internal courtyard system of the complex. At the south, the recreation center creates a buffer zone between the campus and thoroughfare. On the quiet eastern side of the site architects have located the dormitory. On the west, the master courses building and the executive courses building include flat classrooms, auditorium classrooms and boxes. The students will exit the classrooms and

be able to immediately use the boxes to discuss the content of the lessons. The auditorium classrooms are organized with movable chairs and tables to allow professors to circulate among the students.

The curved building plan follows the natural shape of the classrooms. To create an easy way to pass through the buildings each cell touches another. This creates a continuous perimeter that seals the school from the outside. All buildings stand on permeable porticos, and the sequence of columns, transparent rooms and trees becomes clear. Enclosed gardens, courtyards and porticos are a distinctive trait of Milan, one that offers peaceful environments for socializing, studying, gathering and meeting in the open air.

Each cell is organized on a narrow plan, exposing the interior to light and providing large openings onto the park courtyards, optimizing natural ventilation and reducing both cooling and lighting load. Exterior walls are 50% opaque and 50% glazed, providing optimal insulation and reducing cost. Energy strategies will relate the use of water to underground sources or recycled rain water.

The aim of the project is to design a university campus where students, teachers and visitors can be part of a thriving academic community, with an unlimited sense of transparency, closeness to nature, and vision. Architects hope that this flexibility and connectivity will allow the university to progress with unimpeded ambition.

项目名称: A New Urban Campus for Bocconi University
地点: Milan, Italy
建筑师: Kazuyo Sejima, Ryue Nishizawa
用途: dormitory, recreation center, school of management, Bocconi's master and executive building
用地面积: 35,000m²
建筑面积: 17,500m²
造价: EUR 130 million
竣工时间: 2018 (expected)

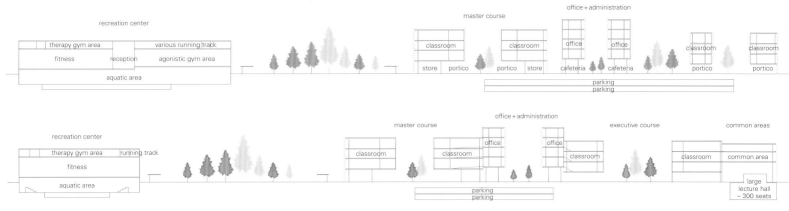

可持续性便利设施 SUSTAINABLE AMENITY

郊野游乐园 _Openfabric + dmau

"郊野游乐园"由Openfabric和dmau事务所共同设计,最近在由理查德·克拉吉塞克基金会和Architectuur Lokaal联合举办的荷兰可持续性游乐场设计竞赛中胜出。该项目位于荷兰莱顿市一个现代化的战后邻里社区。此设计竞赛要求把可持续发展的思想融入到这个体育场地的设计中去。

为此,"郊野游乐园"的设计理念将人造的外侧场地与中央自然的野外景观融为一体,为不同类型的游戏比赛创造了不同的场地环境。外侧场地较正式,可以用来举办体育比赛和有组织的游戏:网球、篮球、五人制足球、60m短跑和跳远。而中央自然的野外景观则鼓励孩子们在树丛中使用像快速生长的柳树这样的自然材料随心所欲地自建和自毁他们自己的游戏空间。这两个不同世界之间建有一座小亭子,用于存放运动器材和建筑材料。一位社区体育负责人负责管理它们。

一条边界性的"丝带"将这两个世界分隔开来,成为把传统游乐场元素与其波浪起伏般的曲线形状融为一体的游乐景观。这一形式参考借鉴了荷兰传统浪漫的景观公园和花园设计,既保护了内部的野外景观,又修建了许多独特的游乐景观,把游乐场内外两个不同的世界连为一体,有攀岩峡谷、既有隧道又有滑梯的小山、带沙滩的池塘、面朝主要运动区的弧形座位区。孩子们在此通过游戏学习怎样穿梭于不同的世界中。维持人造世界和自然世界之间的平衡关系是可持续性发展的精髓,孩子们通过参与游戏和创造性互动可以更好地理解人与自然之间的对话,这是孩子们不可缺少的童年经历,也是现在许多城市孩子们所缺少的经历。

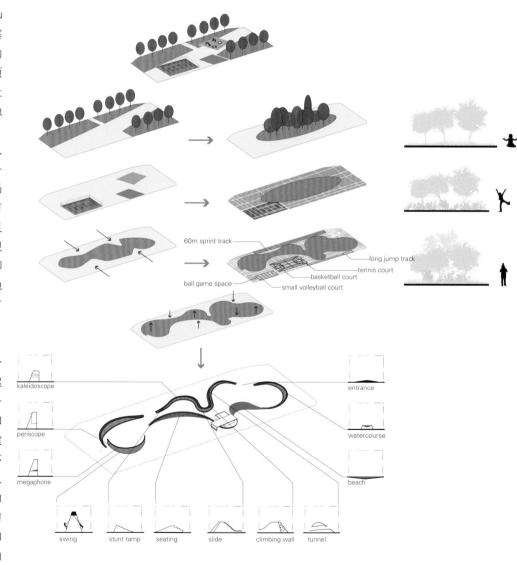

Into the Wild

Into the Wild is a playground design by Openfabric and dmau which recently won the Dutch sustainable playground design competition organized by the Richard Krajicek Foundation and Architectuur Lokaal. The design is located in a modernist postwar neighborhood in Leiden. The competition asked for ideas of how to incorporate sustainable thinking into the design of sports playgrounds.

In response to this the design concept juxtaposes a man-made exterior with a wild natural interior. Each place creates an environment for a different type of play, the formal exterior is a place for sports and structured games: tennis, basketball, 5-a-side football, 60m sprint and the long jump. While in the wild interior, children are encouraged and free to construct and destruct their own play spaces amongst the trees using natural materials such as fast growing willow. A small pavilion sits in-between the two worlds and will act as a storage space for sports and building materials. A neighborhood sports leader will supervise the pavilion and materials.

A boundary "ribbon" separates the two worlds, the "ribbon" becomes a play landscape incorporating traditional playground elements into its undulating and curvilinear form. This form references traditional romantic landscape park and garden design in the Netherlands, it protects the internal wilderness and creates a number of unique playscapes that link the different worlds; a climbing canyon, a hill with tunnels and slides, a pond with a beach, a curved seating stand facing the main sports area. This is the place where children learn through play to navigate between the different worlds. A balanced relationship between the man-made and natural worlds is the essence of sustainability and forming an understanding of this dialogue through participatory play and creative interaction is an essential childhood experience currently missing in many urban areas.

项目名称：Into the Wild
地点：Leiden, Netherlands
建筑师：Openfabric+dmau
项目团队：Daryl Mulvihill, Francesco Garofalo, Barbara Costantino
甲方：Richard Krajicek Foundation
地理定位：52°09'29.45" N 4°27'38.92" E
用途：playground
面积：4,650m²
造价：EUR 450,000
设计时间：2012—(ongoing)

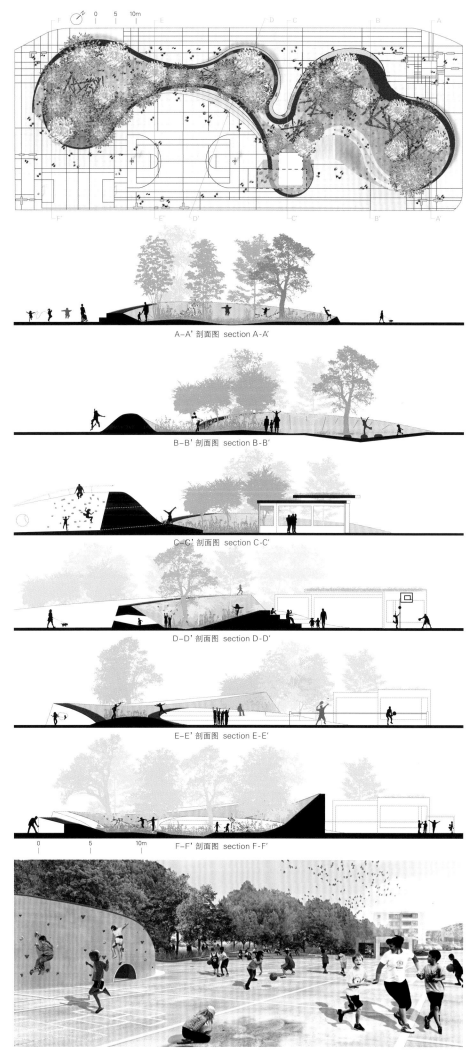

中世纪波斯尼亚历史公园 _Filter Architecture

中世纪波斯尼亚和黑塞哥维那历史公园（泽尼察、波斯尼亚和黑塞哥维那）是本次设计竞赛的名称，目的是以中世纪波斯尼亚历史为主题修建一座主题公园。参与竞赛的建筑师可以自己决定怎样在既定场地中进行布局和选择展现主题的方式，唯一的限制因素是历史公园中必须展示一些重要的历史文物。项目所在地位于城市中心的一块绿地，行人通过的小径密布。建筑师的设计作品是博物馆建筑，他们给它命名为"时空"。

此项目的设计意图是在泽尼察市中心创立一个构成沿波斯尼亚河大Kamberović公园不可缺少的组成部分的展览空间，设计理念基于历史事件的确定性，即对历史前因后果的认识，同时避免关联过往历史所带来的伤感情绪和只人为虚假表现该国历史的一部分。建筑空间致力于建立起与游客情感上的沟通。建筑入口以单行道方式逐渐消失于地下，通道两侧的墙以镜子贴面，并最终回到起点，但在不同的水平面上，创建一种相对空间感，揭示了上文所提到的"时间"理念。

建筑的结构体系由25块完全相同的混凝土模块组成，模块与模块组合在一起，形成螺旋几何形状。预制混凝土模块形状如"盒子"，排列成圆形，并在垂直方向上相互错开。

表面处理

外部：面朝外部公园的螺旋形结构表面镶嵌着感光玻璃，使能够接受阳光照射的那一侧形成反射。因此，白天时建筑物反射着公园的满园绿色；夜晚时，当打开安装在玻璃后面的照明开关，建筑就变成一个绚丽夺目的"巨大发光体"。螺旋形结构面向内部广场的一面覆有白色混凝土，表明公园的这一部分属于更加私密的环境。

内部：博物馆内部以镜子饰面，镜面上印有所展览的历史文物，目的在于营造一种无限空间的错觉。在镜子中，游客的身影和历史文物的影像交织在一起，形成独特的视觉体验。

设计和建造难题

由于展馆紧挨着公园的一条主路，因此展览空间的形状呈流畅的线性，与行人的移动方向保持一致。路径被设计成弯曲状以形成一个广场，该广场由展馆围合起来，并逐渐形成一个分段式的螺旋结构，缓缓融入公园的自然形态之中。

开发该展馆的总预算是500 000欧元，因此就要求建筑师能拿出一个结构性的解决方案来降低造价，最终决定使用便宜的预制混凝土模块，这样展馆建设又快又廉价。

Historical Park of Medieval Bosnia

The name of the competition is Historic Park of Medieval Bosnia and Herzegovina (Zenica, Bosnia and Herzegovina). The intention of this competition was to create a thematic park, with medieval Bosnian history as the subject. It was left to the competitors to decide how to organize the given site and to choose the approach. The only limiting factor was a set of the most important historic artifacts, which had to be presented at the place. The chosen site was a green area in the city center, well provided with the pedestrian paths. Architects' design result was a museum pavilion, which they named "timespace".

The design was prompted by the idea of an exhibition space forming an integral point of the large Kamberović Park alongside the River Bosna, in the center of the town of Zenica. The concept was based on a deterministic approach to history – as a series of causes and consequences, while avoiding falling into a trap of a pathos-ridden and artificial representation of a part of their national history. The architectural space aspires to create an emotional communication with the user. The entry

sequence – disappearing underground as a one-way movement between walls lined with mirrors and then returning to the beginning – at the same point but on a different level, creates a sense of relative space and reveals the time as the content.

Structural system consists of 25 identical concrete modules, which combined together create a geometry of spiral. Prefabricated modules are shaped as "box-like" segments of the circle that slip between each other in vertical sense.

Surface treatment

Exterior: The part of the spiral facing the exterior park is coated with photo-sensitive glass, that creates reflection on the side which is more exposed to the light source. Therefore, building reflects the green of the park during the day, and serves as a "giant glow" during the night, when the illumination placed behind the glass is switched on. The part of the spiral facing the inner square is coated with white concrete, referring to the fact that this part of the park serves as more private ambient.

Interior: The interior walls are covered with the mirrors which have exhibition artifacts printed on their surface. Their purpose is to create an illusion of indefinite space, inside which the image of the visitor and the image of an artifact are being combined together into a unique visual experience.

Issues faced during design and set up

Due to the fact that the pavilion is placed next to the major park trail, the shape of the exhibition space was determined to be a line of transit that fluently matches the pedestrian flow. The line curves in order to form a square that is surrounded by the pavilion and grows into a segmented spiral that smoothly interpolates into a park morphology.

Total amount of money assigned for the development of the pavilion is EUR 500,000. Therefore, it was necessary to find a structural solution that manages to lower the price. The decision was made to create a cheap system of prefabricated concrete modules that enable the pavilion to be built quickly and inexpensively.

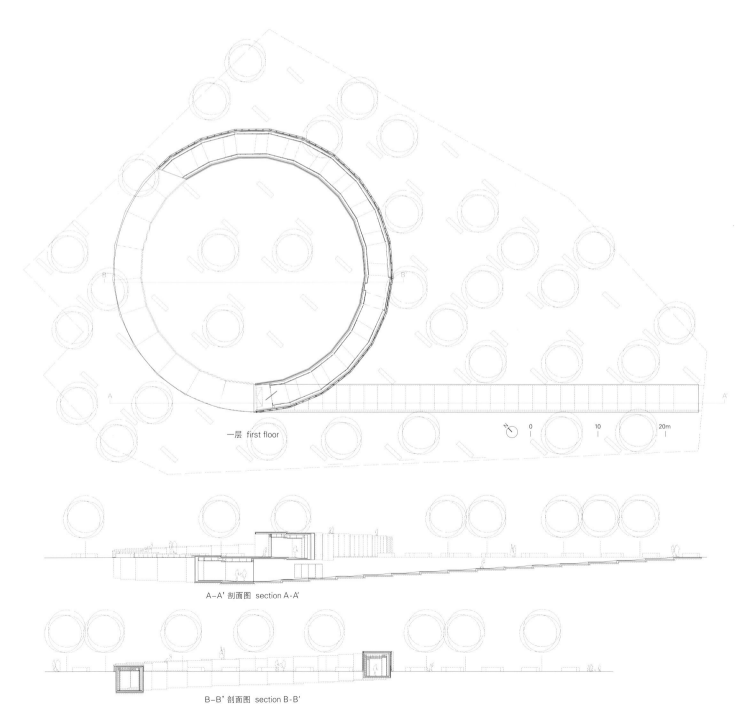

一层 first floor

A-A' 剖面图 section A-A'

B-B' 剖面图 section B-B'

项目名称：Historical Park of Medieval Bosnia
地点：City of Zenica, Bosnia and Herzegovina
建筑师：Filter Architecture
甲方：Government of the Zeničko-Dobojski Canton
设计时间：2009—2011

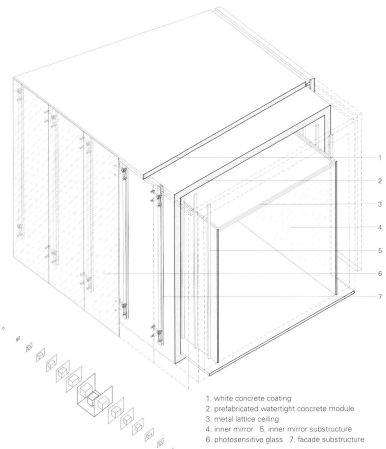

1. white concrete coating
2. prefabricated watertight concrete module
3. metal lattice ceiling
4. inner mirror   5. inner mirror substructure
6. photosensitive glass   7. facade substructure

镜子实验室2.1 _VAV Architects

"镜子实验室"的设计理念是设计出能不断改变其周围环境的装置。

项目场地位于加泰罗尼亚一个小镇市郊的一座古桥内,是一处宁静、安详、平静之地。结果证明,这项设计既有趣又有挑战性。

建筑师的目的是通过最低限度与原有景观并列、不喧宾夺主地将镜子嵌入到原有的景观中,从而达到锦上添花的效果。在古桥的拱内简单地嵌入镜子,就在原有景观上增加了一个全新的、激动人心的维度。在拱内,镜子成功地复制了拱内外的景色,把空间倒置;在拱外,镜子又能捕捉和映现外部风景。从拱内穿过,游客可以与镜子装置互动、融合,沉浸于真实与影像之中。

"镜子实验室"向游客们展现了一个变化莫测的奇妙仙境,原本固定不动的景观像变色龙一样变化多端,站在原地不动就可观赏到变换的风景使游客在此可以体验畅游仙境的愉悦。因此,"通过镜子"看世界,看到的是一个离奇的不真实的世界,使我们可以领略周围千变万化的美景,领略其别样的美。

Mirror Lab 2.1

The conceptual thought behind "Mirror Lab" was the creation of device with the capability of constantly changing and altering its environment.

The site – an old bridge on the outskirts of a small Catalonian town – is a place of tranquil serenity and calmness, proved interestingly challenging to work with.

Architects' aim was through minimum juxtaposition, a humble insertion to enhance the scene through non-obtrusive interaction with which already existed. With a simple insertion into the arch of the bridge, the mirror adds a completely new and exciting dimension. Inside the arch, it successfully replicates the views and inverts the space while outside it captures and frames them. Journeying through the archway the visitor can interact and fuse with the installation and immerse themselves in both what's real and merely reflected.

"Mirror Lab" affords a Wonderland of ever changing scenarios whereby that which is fixed becomes "Chameleon Like" and enables the viewer to experience the joys of journeying through the changing scenes without moving location. Thus as we view "Through the Looking Glass" we are continuously seeing a surreal world in our eyes, and we can savour the ever evolving beauty of the landscape around us.

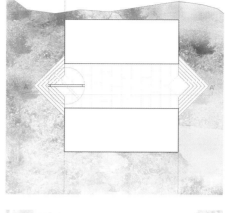

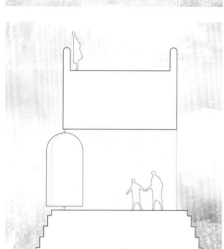

A-A' 剖面图 section A-A'

   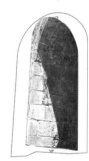 

项目名称：Mirror Lab 2.1
地点：Bridge de Sant Roc, Olot, Spain
建筑师：Pablo Bolinches Vidal, Darragh Breathnach, Daria Leikina
竣工时间：2011
摄影师：Miquel Merce (courtesy of the architect)

格兰富大学学生宿舍 _CEBRA

格兰富大学学生宿舍楼引人注目的特点之一是其位于奥尔胡斯新港口规划区的核心位置,也就是说,在引人注目之地建了一座低成本的住宅建筑。奥尔胡斯港以前是个集装箱港口,如同许多其他工业港口一样,奥尔胡斯港现在正发展成为一个富有活力的新区。格兰富大学将是本地区首批竣工的项目之一,竣工之后,将有7000名居民在此安家,还将包括12 000个工作场所,总占地面积达800 000m²,将成为欧洲最大的城市港口规划项目之一。

此建筑主要是为在读的大学生提供住处。为了突出这一主要功能,建筑外观设计成了垂直条纹,暗喻书籍。从远处看,建筑就像书架上摆放的书;在近处看,就像浓缩的微型曼哈顿,塔楼林立。每一栋小塔楼里外都做了不同的处理,使用了不同的材料,窗户也设计得形式多样,创造了多种多样的居住空间——有适合单身居住的,有适合夫妻居住的,也有适合朋友一起居住的。为了楼顶部技术设备的安装、强调主入口并在屋顶建一个鸟瞰整个海港和城市的观景平台,每座塔楼设计得高度不一。这些设计完全出于功能性考虑,却赋予了这个微型曼哈顿属于自己的独特天际线。

塔楼位于四个角上,形成洞开的内部空间,被设计成集功能和美观于一体的中庭,壮观且引人入胜。每套学生公寓都通过环绕中庭的走廊进入,共12层。走廊面向中庭的墙面全部镶嵌了镜子,里面人影攒动,反射出犹如万花筒一样千变万化的世界,使相对狭窄的中庭空间感无限扩大。

大学生宿舍楼也是一个社会交往的空间,镜子的使用使这一公共空间更具活力。镜子的反射可以帮助你隔着几层楼定位,从一个位置,你可以看到楼上或楼下的人或事。同样,中庭丰富多彩的颜色也有助于人们隔着几

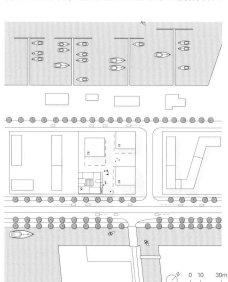

层楼定位。建筑物的立面由几个塔楼组成,与此相对,公寓门的形状及其不同色彩形成红色、橙色和黄色的垂直彩带,从一楼一直延伸到顶楼。

镜子不仅是创造引人注目的空间感的一种成本低廉的方法,也能通过反射自然光减少人工照明的需要。镜子最大限度地反射来自中庭上部大天窗的阳光,使阳光从顶部一直倾泻到一楼大厅。

Grundfos College Student Dormitory

One of Grundfos College's remarkable characteristics is providing student housings in the heart of Aarhus' new harbor front development – that is, a low cost residential building on a very attractive site. Like many other industrial harbor fronts, the former container port of Aarhus is being transformed into a dynamic new neighborhood. Grundfos College is one of the first finished projects in the area that on completion will be home to 7,000 inhabitants and provide 12,000 workplaces. Its total site area amounts to 800,000m² making this development one of Europe's largest harbor front city developments.

To reinforce the primary purpose of the building, which is to provide a place for students to live while they study, the project works with vertical stripes as a metaphor for books. From afar this makes the building look like books on a shelf, while close up it resembles a condensed micro Manhattan with small bundles of mini towers. Each little tower is treated differently both inside and out with different materials and window openings. This creates a variety of living accommodation, suitable for singles, couples or friends living together. The towers are designed to be different heights in order to incorporate technical facilities on the roof, accentuate the main entrance and provide a roof top terrace with a view over the waterfront and the city. These purely functional considerations end up giving this micro Manhattan a unique skyline of its own. The "towers" are placed on the perimeter of the site establishing an open inner space that uses a simple design strategy to create an atrium with attractive functional as well as spectacular aesthetic qualities. The individual apartments are reached from balconies, which are encasing the 12-story atrium with mirror clad balcony fronts. The mirrors are transforming the sense of space by expanding the relatively narrow atrium with endless kaleidoscopic reflections of itself and the people moving through it.

The mirrors also contribute to a more activating common space by supporting the social aspects of a dormitory building. The reflections assist orientation across several floors from one location allowing you to see people or activities that are located directly beneath or above you. In a similar way, the colors in the atrium enhance orientation across floors. As an equivalent to the facade's division into towers, the graphic and apartment door colors are used to form vertical bands of red, orange and yellow from the bottom to the top floor.

Besides being a low cost method to create striking spatial features, the mirrors also reduce the need for artificial lighting by reflecting a maximum of daylight from the large skylight all the way down to the lobby.

项目名称:Grundfos College Student Dormitory
地点:Grethe Løchtes Gade, Aarhus, Denmark
建筑师:CEBRA
合作者:Sjælsø, Niras
甲方:Engineering College of Aarhus
用途:student apartments, dormitory
面积:6,000m² / 施工时间:2010—2012

墙体工艺
Wall Graft

在现今欧洲建筑办公室设计的普通作品里，人们很少能看到构建在白板上的宏伟建筑，这不是因为资金的限制，而是"建筑遗产"的概念被延伸了："建筑遗产"现在不仅包括一些单座纪念建筑和文化作品，还包括更多的平庸之作。时间似乎仍然是某些元素是否值得保留的决定性因素。在其保留的原则上，有一种道德标准也因此获得了更多的认可，即"古老的"就是好的，而"新型的"是可以摒弃的。

因此，目前，欧洲建筑师们受到的这类委托的重要部分即是对场地进行设计，对这里过去的废弃建筑进行修葺，在这些废弃建筑中，废墟是比较特殊且特别珍贵的。人们都倾向于喜欢废墟，因为废墟能使人们与过去的有关事物产生共鸣。对废墟的这种普遍的热情也感染了知识分子和建筑师们，废墟是一种位于废弃场所的某种事物的象征，处于艺术和自然之间，是人工与有机形成的混合结构。

修复过去的废弃建筑对建筑师而言是一种挑战，因为他们不得不给那些作品再加上"一层"情感光环。当代的建筑师们采用很多方法来应对这一挑战，比如，废墟可能承担一个表述性的角色、一个纪念的符号或者某一个确定的功能。然而，在这些最成功的建筑实例中，一些结构的象征和形式展示了精确的但又不失批判性的评估，这种评估远远地超过了古迹的商业价值。在这些案例里，新加的"一层"光环成为对废墟美学方面的混合运用所进行的礼赞。

The grand architectural gesture built over a tabula rasa is now seldom spotted in the ordinary work of an European architectural office. This is not due to economic constraints, but because of a broadening notion of architectural heritage, which now includes not only singular monuments and highbrow architectural pieces but also more prosaic structures, buildings or ensembles. However, time seems to be still the determining factor that validates whether something deserves preservation or not. Concerning the criteria for preservation, the following moral assessment is thus gaining momentum: If it's old, it is good, if it's new, we can discard. Hence, nowadays, an important part of the commissions received by European architects deals with designing projects for sites where the derelicts of the past have to be revamped. Ruins are a special and very cherished type of those derelicts. And people tend to like ruins, as they often resonate with something that creates a link with the past. This popular fascination with ruins also pervades the intellectuals and the architects among them. Ruins are a token of something that stands somewhere in a limbo, between art and nature. They are hybrid structures that were created not only by the hands of men but also organically generated. Revamping those derelicts of the past thus creates an additional challenge for architects, as they have to add another layer to an artefact that has already a charged aura. Contemporary architectural approaches show many different ways in which the design approach copes with this challenge. For example, the ruins can assume a performative role, a commemorative symbol or a deterministic function. Some of the most successful cases, however, reveal a shrewd critical assessment of the symbolic and formal qualities of those structures, that goes beyond a mere commodification of their antiquity. In these cases, the new layer thus becomes a celebration of the aesthetic hybrid of ruins.

柯尔顿公寓/3ndy Studio
塔利亚剧院/Gonçalo Byrne Arquitectos + Barbas Lopes Arquitectos
Potxonea住宅/OS3 Arkitektura
修复过去的废弃建筑：对混合结构的颂扬/Nelson Mota

Corten Apartments/3ndy Studio
Thalia Theater/Gonçalo Byrne Arquitectos + Barbas Lopes Arquitectos
Casa Potxonea/OS3 Arkitektura
Revamping the Derelicts of the Past: In Praise of the Hybrid/Nelson Mota

《L' Adoration des bergers》，作者Nicolas Poussin，伦敦：国家美术馆，1633—1634年
L'Adoration des bergers by Nicolas Poussin, London: National Gallery, 1633~1634

《Le antichità Romane, t. 1, tav. XXXII》，作者Giovanni Battista Piranesi, 1756年
Le antichità Romane, t. 1, tav. XXXII by Giovanni Battista Piranesi, 1756

修复过去的废弃建筑：对混合结构的颂扬

文艺复兴精神培养了建筑师对废墟的热情。从15世纪开始，人们便对探寻历史，以对现在更加了解展示出兴趣，这引发了人们从学术和情感方面对废墟的颂扬，将其作为与过去的链接，这将深刻地影响文学和绘画创作。在绘画方面，以公元17世纪的法国古典主义画家Nicolas Poussin为例，他一生中的大部分时间都住在罗马，其大部分作品都与各种废墟有关，而这些废墟经常与之前某一时期的衰落产生共鸣，如新世纪兴起的基督教取代了罗马帝国的异教。对Poussin来说，废墟不只是一处实际或是逼真的场景，它还是传递宗教和政治信息的工具，我们从他的1633年的作品《牧羊人对耶稣出生地的崇拜》中可以看出Poussin对废墟的看法。一个世纪之后，浪漫主义的出现使我们对废墟的兴趣达到前从未有的程度。这一次，废墟被推崇成可以入画的元素，这些元素完全可以描绘那些浪漫多情和虚幻世界中的抽象事物。例如18世纪废墟应用于园艺和景观设计中，在那时的许多英国、法国和德国的公园中，如果没有真正的废墟，就仿造一些废墟。此外，废墟也被作为说明历史的更加真实的证明。

事实上，从18世纪Piranesi在罗马绘制的著名蚀刻版画中，我们可以看到废墟代表了实际和情感因素的最佳结合，在他的作品《罗马古事记》中，Piranesi不只是从考古学的角度展现了罗马废墟，他还表现了日常的场景，在这些场景中，"过去的片段"被人类和自然所"征服"。Piranesi对建筑文化的影响有所记载，并且增加了建筑师们对废墟的热情。譬如说，在19世纪，对那些接受"罗马大奖"的建筑师来说，描绘罗马帝国的废墟成为他们的日常工作。另外一家致力于在地中海盆地探索西方文明摇篮的、被称作"伟大旅行"的建筑机构，也对增加人们对废弃物的兴趣起到了作用。当年轻的Charles-Edouard Jeanneret，也就是后来的著名建筑大师勒•柯布西耶，在1911年去参观雅典卫城（其东方之旅中的一站）时，帕台农神殿的废墟给他留下了深刻的印象，"他连续三个星期每天都要参观帕台农神殿，画草图或者拍照，甚至把神庙比作一台机器。"[1]那幅著名的Charles-Edouard站在拆除的多利安柱旁边的绘画就证明了建筑和废墟之间的紧密联系。

现在，这种关系变得更加实际，并且我建议减少其浪漫的成分。随着建筑遗产增值过程的加速，废墟越来越成为建筑师每天工作中要

Revamping the Derelicts of the Past: In Praise of the Hybrid

The spirit of the Renaissance fostered architect's fascination with ruins. From the 15th century on, the interest in understanding the present by rediscovering history triggered an intellectual and emotional appraisal of the ruins as a link with the past, which would deeply influence literary and pictorial creations. In painting, for example, many works of the famous 17th century French classicist painter Nicolas Poussin, who spent most of his life living in Rome, are impregnated with depictions of ruins, which often resonate with the decline of a former status-quo, i.e. the paganism of the Roman empire, superseded by a new age that was emerging, Christianity. For Poussin, ruins were thus instrumental to convey a religious and political message, more than an actual or verisimilar setting, as can be seen in his 1633 depiction of the *Nativity Scene of the Adoration of the Shepherds*. One century later, the emergence of Romanticism would foster the interest in ruins to an extent than was never seen before. This time, however, ruins were championed as "picturesque" elements that would be thoroughly used for their romantic qualities and their abstract representation of an imaginary world. In 18th century gardening and landscaping, for example, this was taken to such an extent that fake ruins were created when no genuine ones existed, as was in the case of many English, French and German parks. Yet, ruins were also used as a more factual support to document the past.

In fact, an excellent conflation of both factual and emotional representations of ruins can be seen in Piranesi's famous etchings of Rome, made in the 18th century. In his *Antichita Romanae (Roman Antiquities)* Piranesi goes beyond a mere archaeological representation of Roman ruins and also shows everyday scenes, where those fragments of the past were "conquered" by both humans and nature. The influence of Piranesi in architecture culture is well documented, and contributed to increase architect's pleasure of ruins. For example, in the 19th century, mapping the ruins of the Roman Empire became a routine activity for the architects who received the Prix de Rome. Another architectural institution, the so-called Grand Tour, an trip to discover the cradle of western civilization in the Mediterranean basin, also contributed to foster architect's attraction with the derelict. When the young Charles-Edouard Jeanneret, who later became better known as Le Corbusier, visited the Acropolis in Athens in 1911 as a stop in his "Voyage d'Orient(Journey to the East)", the ruins of the Parthenon caused such an impression to him that *"he revisited the site every day for three weeks, sketching and photographing, even comparing the temple to a machine."*[1] The famous picture of Charles-Edouard standing next to a dismantled Doric column testifies to this pas-

塔利亚剧院,从南面望去
Thalia Theater, seen from the south

Potxonea住宅,位于两个广场之间的街区
Casa Potxonea, located at the block mediating two squares

应对的材料,这不仅仅是因为废墟能够昭示远古时期的荣誉,比方说庙宇或宫殿,还由于那些常见的不太古老的遗迹也是建筑师必须面对的材料。然而这给建筑学科带来了新的挑战,正如上面所提到的,在18世纪,如果没有真正的废墟,人们会仿造废墟来满足浪漫的怀想。今天,建筑师们在一些建筑手法中采用一种相似而又截然相反的方法,特别是处理位于历史遗产保护地的建筑时。建筑或场地所带有的象征意义和文化价值经常会对建筑师或者公共机构设计一些所谓的前卫建筑起到阻碍的作用。作为一种全民意识,通常的结果是仿造各种废墟,诸如对其进行类似的延伸、原本保留或者原地保护。幸运的是,所有这些都不包括在我们以后要讨论的建筑中,因为我们所要讨论的建筑采用截然不同的方法来应对挑战——它们把过去与现在结合在一起。

由Gonçalo Byrne建筑事务所和Barbas Lopes建筑事务所组成的团队设计的里斯本塔利亚剧院是把废弃物和新建筑结合的大师级作品。这座于1843年竣工的新古典主义风格的小型建筑存在的时间很短,在建成后19年便毁于大火而成为废墟,该废墟为以后建筑的批判性重建提供了指导方针。在重建中,废墟与新的建筑元素似乎在捉迷藏。在外部,建筑师重建了新古典主义风格的立面、门廊和门厅,与整体体量(突出了观众席和舞台)形成了强烈的对比。为了使这一体量内的互动更具有戏剧性,带有增设设施的地面玻璃结构将主体建筑与城市环境连接起来,不但形成了一个临街立面,而且还在剧院后方形成一个露台,以俯视城市动物园的邻近公园中栽种的异国植被。在观众席和舞台区,赤褐色的混凝土外壳覆盖了由1862年大火在观众席和舞台上造成中空空间。在这些区域的内部,原有的墙体遗迹以混凝土壳模板的形式被保留下来。据建筑师所说,由于其本身就是一场精彩的演出,因此这些墙体得到了精心的清理和保护。

由3ndy工作室设计的、将一座19世纪位于威尼斯省维戈诺地区的宫殿改造成的公寓,采用不同的方式来达成新旧之间的交流。如果说塔利亚剧院更加注重建筑的构造轨迹,那么3ndy工作室设计的柯尔顿公寓则注重其布景,以呈现更加卓越的效果。在该设计中,建筑

sionate relation between architects and ruins. Nowadays, this relation grew to something more tangible and, I would suggest, less romantic. With the increasing valorisation of architectural heritage, ruins are now more and more an everyday material that architects have to cope with in their works. Not only the ruins that testify to the glory of ancient times, such as temples or palaces, but also the remnants of the not-so-old architecture of the everyday. However, this brings about new challenges to the architecture discipline. As it was mentioned above, in the 18th century, the romantic appeal of ruins was such that fake ruins were created where no genuine ones could be found. Today, a similar yet contrasting approach can be seen in some architectural approaches, especially those that deal with buildings in heritage protected sites. The symbolic and cultural value attached to those buildings or sites, often hinders architects and public authorities from designing or approving so-called bold architectural actions. As a kind of populist consensus, this frequently results in the creation of all sorts of fake ruins, such as mimetic continuities, fundamentalist preservation, or immobilizing safeguarding. Fortunately, none of this happened with the projects discussed below, as they illustrate different approaches that cope with the challenges brought about by mingling the past with the present. The project for the Thalia Theater in Lisbon, designed by the team Gonçalo Byrne Arquitectos + Barbas Lopes Arquitectos, reveals a masterly done conflation of the derelict and the new. The ruins of the short-lived small neoclassical theater, destroyed by a fire only nineteen years after its completion in 1843, generated the guidelines for a critical reconstruction of the building. In this reconstruction, the ruins play hide-and-seek with the new elements. In the outside, the remnants of the neoclassical facade, portico and foyer were reconstructed and powerfully contrasted with the monolithic volumes that extrude the plan of the audience and the stage. To add more drama to this volumetric interplay, a ground glazed structure with additional facilities articulates the main building with its urban situation, thus creating both a street facade and a patio at the back of the theater, overlooking the exotic vegetation of the neighbouring park of the city zoo. A terracotta concrete shell covers the void created by the 1862 fire in the audience and stage areas. In the interior of these areas, the ruins of the original walls were kept as a sort of formwork for the concrete shell. They were thus cleaned and preserved with the deliberate purpose of, according to the architects, being performers in a spectacle of their own.

The project designed by 3ndy Studio for the transformation into apartments of the ruins of a 19th century palace in Vigonovo, in the province of Venice, shows a different approach to the interplay between the old and new. If in the case of the Thalia Theater the focus was more on the tectonic locus of architecture, in 3ndy Stu-

1. William J. R. Curtis, "The Classical Ideas of Le Corbusier", The Architectural Review 230, no.1376, 2011 October, p.32.
2. Paul Zucker, "Ruins. An Aesthetic Hybrid", The Journal of Aesthetics and Art Criticism 20, no.2, 1961 December, p.119.

师刻意地追求一种和尚未损毁的废墟之间轻松相处的且具有纪念意义的关系，而拒绝单纯的重建。他们认为单纯的重建是一个必然的错误。举例来说，我们可以从建筑物的立面中看到他们的设计方式。T形体量采用了彩色粉饰、砌砖还有柯尔顿钢来作为其不同的饰面。柯尔顿钢是整座建筑的特点，建筑以此命名。它应用在整个损毁宫殿立面的重建中，这是艺术家Giorgio Milani的杰作，使部分建筑焕然一新，是将废墟之美转化为新的重要装置。

微妙的平衡允许建筑新旧元素之间的调节，对建筑师来说是一个主要的挑战。位于西班牙乌苏尔维尔市的Potxonea住宅是由OS3建筑事务所设计的，以达到项目旨在传递的一种设计平衡，该设计是由经过深思熟虑的、委托限制所凝聚的精华发展起来的，这些委托限制本身就是在过去与现在之间获得的平衡。该建筑场地位于城市中心，即一个街区的最顶端，两侧是风格迥异的两个广场，毗邻一座主教堂，因此其位置在设计理念中发挥着主要的作用。

在这个案例中，现存房子的废墟要保留，因其立面是朝着城市最具代表性的广场——Dema广场，这个废墟是产生最后设计方案的根本元素。其立面又附加了一个新结构，这个新结构在面向邻近建筑的方面展现了其不同之处，这种不同在其具体化以及实体与体量之间的关系方面是可以感知的，关于项目限制方面的建筑探讨最终可以完全显示建筑师的能力，即将看似矛盾的元素结合在一起，形成一个协调但是混合型的整体。

德国建筑师和艺术历史学家Paul Zucker认为，这种混合性，实际上与废墟之美产生了共鸣。1961年，他认为废墟是具有美学价值的混合体。他说道，"这些废墟无论是时间还是人为摧毁的，都是不完整的，他们代表着一种人工形式与有机自然的结合。所以废墟的情感层面是模棱两可的：我们说不清废墟是属于艺术之美还是自然之美。"²因此，以上所述的建筑项目展示了建筑师处理这种"模棱两可"的几种方法，但他们有一个共同的特点，就是摆脱了废墟的"多愁善感"（按照Zucker的叫法），以对采用当代建筑方法所展现的过去重新进行批判性的评估。

dio's Corten Apartments is the scenographic aspect that assumes greater prominence. In this case, the architects deliberately pursued a playful and commemorative relation with the extant ruins, thus rejecting a mere reconstruction of the building, an attitude deemed by them as inevitably false. This approach can be seen, for example, in the facades of the building. The T-shaped volume receives such different finishings as coloured stucco, brickwork masonry and corten steel. The later material gained special character, as the name of the project testifies. It was used in the reconstruction of the facade of the ruined palace, done by the artist Giorgio Milani, thus dressing a part of the building and creating an important device to negotiate the inclusion of the aesthetic of the ruins into the new.

The delicate balance that allows the architectural accommodation of the new into the old represents a major challenge to architects. In the case of the Casa Potxonea, designed by OS3 Arkitektura in Usurbil, to achieve that balance the project aims to deliver a design shaped by a thoughtfully distillation of the commission constraints, which are themselves an outcome of the accommodation of the past into the present. The location of the building site, in the city center at the top of one block mediating two squares with completely different character, and next to the main church, thus played a fundamental role in the design concept.

In this case, the ruins of an existing house, whose facade towards the city's most representative square, the Dema Square, should be preserved, became the fundamental element to produce the final solution. This facade generates a new structure attached to it, which is keen in expressing its difference towards the neighbouring buildings, a difference that can be perceived in its materialization or in the relation between solid and voids. The architectural negotiation of the project's restraints delivered an outcome where the architects fully demonstrate their ability to bring together elements seemingly conflictive, into a coherent yet hybrid body. This hybridism, in fact, resonates with the aesthetic realm of ruins, according to the German architect and art historian Paul Zucker. In 1961, he argued that ruins were an aesthetic hybrid. "*Devastated by time or wilful destruction, incomplete as they are*", Zucker argued, "*they represent a combination of man-made forms and organic nature. Thus the emotional impact of ruins is ambiguous: we cannot say whether they belong aesthetically in the realm of art or in the realm of nature.*"² Hence, the projects discussed above show different approaches on how architects deal with this ambiguity, but they share a common ground, though. They go beyond the "lacrimose sentimentality" of the ruins, as Zucker called it, thus contributing for a critical reassessment of the presence of the past in contemporary architectural approaches. Nelson Mota

柯尔顿公寓
3ndy Studio

二层 second floor

一层 first floor

  3ndy工作室承担了重修Campiello"柯尔顿公寓"的设计委托，该建筑于2011年9月竣工，完成了其本身成为意大利威尼斯维戈诺地区重要的一部分的目标，成为一处新的社会聚集场所和一处历史中心。

  在2009年，当3ndy工作室接受修复宫殿的委托时，这座建筑已被遗弃，且破败不堪，但是约20年前的一场大火并没有对其造成主要的破坏，而是由于随意的整修，如同失败的整容手术，完全损毁了建筑立面优雅的装饰元素。

  恢复这座古老建筑的生命，找回古时的荣耀意味着召回房屋里曾经生活的"声音"——历史的碎片、日常生活的点点滴滴以及传统，所有这些都能在一张纸找到并且探寻到踪迹，这张大纸最后成为Campiello项目中由190块柯尔顿钢组成的大型雕塑，这些钢铺开后的面积为300m²，可以被解读成一本书中大型且神秘的一页。

  这个设计理念旨在再现19世纪毁于火灾的建筑设计。重建这样一个建筑物可能是一个机遇，但是也可能会是个错误，所以建筑师决定采用"纪念"的设计手法。

  由3ndy工作室设计的立面复制了原大小的损毁立面，但是却采用了灰泥来表现时间点。为了能完成这个立面设计，使其独一无二，3ndy工作室在Philippe Daverio的建议下，邀请了皮亚琴察市的雕塑家Giorgio Milani参与到其艺术创作中。

  Milani在"这页书"上进行书写，在柯尔顿钢上雕刻了许多单词的字母，使"这页书"被字母所产生的"满"和"虚"所丰富。这个字母作品"记忆中脚步的回声"诞生在这里，组成这个作品的字母和符号都是由艺术家Giorgio Milani根据其公元19世纪的起源（19世纪是原建筑落成的时间）及其美学外观来精心挑选的。22个不同的大小写字母的字体也有所不同，布局协调，使其美学意境和文化蕴含浑然一体。

Corten Apartments

3ndy Studio was committed for the renovation of the Campiello "Corten Apartments", which has been inaugurated on September 2011. The project reached its goal of retraining a very important part of Vigonovo Venice, realizing a new place of social aggregation in completion with the historical center.

In 2009 when 3ndy Studio was commissioned to renovate the Palace, it appeared abandoned and dilapidated, but no major damage had been caused by the fire that had hit him about twenty years before, but by the reckless restoration which, as a botched cosmetic surgery, had totally distorted and stripped its elegant decorative elements on the facade.

Taking a place back to life and bringing it back to its ancient glory means to recall the voices that lived in it, the history fragments, the pieces of daily life and tradition that can be found and traced out on a sheet of paper. This huge piece of paper has become, in the project of the Campiello, a big sculpture made of 190 sheets of corten steel which spread for 300m² and are to be read as a giant and enigmatic page of a book.

The idea is to present again the design of the nineteenth-century building, destroyed by a fire. Performing a reconstruction of this building might have been a chance but it would have been false, so the architects decided to take the way of commemoration.

The facade designed by 3ndy Studio reproduces the dimension of the destroy one, but suggests the spot of time on the plaster. In order to complete the facade and make it an exclusive one, 3ndy Studio under the suggestion of Philippe Daverio, asked Giorgio Milani, a sculptor from Piacenza, to take part in this big artistic operation.

Milani wrote on this page, enriching by using fullness and emptiness due to the words of the alphabet carved from the panels of corten steel. From here has been born *"Footfalls Echo in the Memory"*, a work of lettering, which means composing letters and symbols carefully chosen by the artist, Giorgio Milani, according to their nineteenth-century origin (the age in which this building has been built) and their aesthetics dress, balancing 22 different alphabets, upper-case and lower-case letters, of different fonts in order to have an aesthetic and cultural harmony. 3ndy Studio

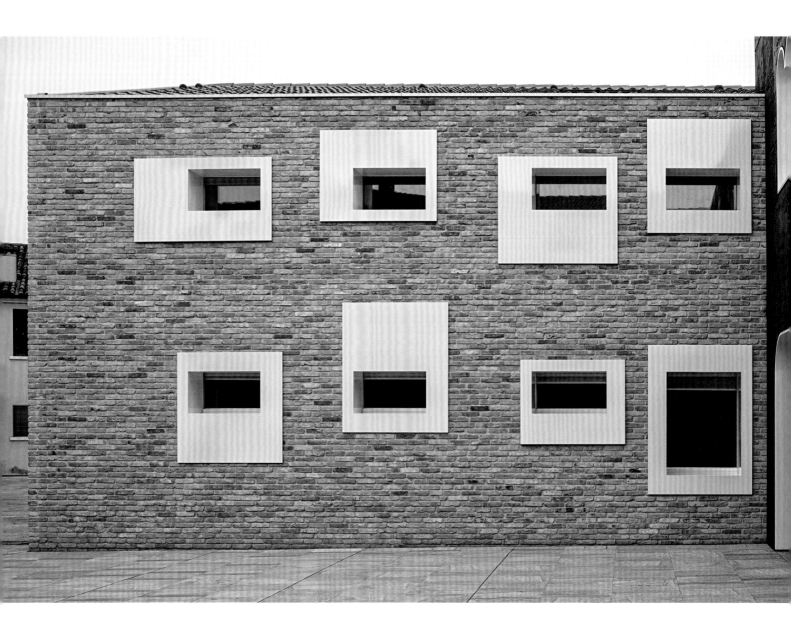

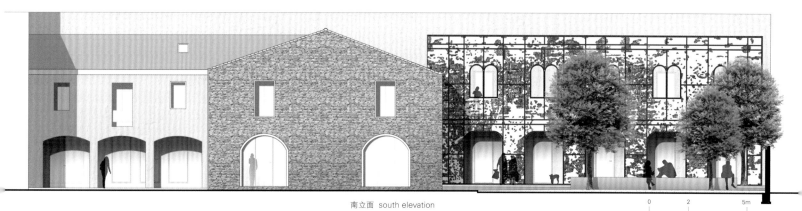

南立面 south elevation

30

原始立面 1800	损毁过程 1980	大火 1985	概念 2009	竣工 2011
origin 1800	degradation process 1980	fire 1985	concept 2009	completion 2011

5 十四行诗《126 Amor》,作者Lope de Vega(译成意大利文)

5. Sonnet *126 Amor* by Lope de Vega (translated into Italian)

13~14 《四首四重奏》,作者T.S. Eliot (译成意大利文)

13-14. *Four Quartets* by T.S. Eliot (translated into Italian)

10 《看不见的城市》,作者Italo Calvino

10. *Invisible Cities(Le Città Invisibili)* by Italo Calvino

12 弗朗切斯科·博纳米的语言

12. Francesco Bonami's words

Giorgio Milani对立面进行设计
Giorgio Milani working on the facade design

在柯尔顿钢板上雕刻字母
carving the letters on the corten steel plate

Texts perforated on the wall, translated in English

1-7. *Sonnet 126* **by Lope de Vega**
To faint, to dare, to rage
be brutal, tender, generous, equivocal
buoyant, mortal, defunct, alive
faithful, traitorous, cowardly and brave
To find outside of goodness, no peace or respite,
appear ecstatic, melancholy, modest, haughty
irate, courageous, on the run,
satisfied, offended, suspicious
To turn your face away from plain disillusion,
drink poison that is smooth liquor,
overlook profit, love pain
Believe that heaven is inside hell,
give life and soul for disillusion,
this is love, whoever has tasted knows.

8-9. *Burnt Norton* **by T. S. Eliot**
Words strain,
Crack and sometimes break, under the burden,
Under the tension, slip, slide, perish,
Decay with imprecision, will not stay in place,
Will not stay still.

10. *Invisible Cities* **by Italo Calvino**
At Melania, every time you enter the square, you find yourself
caught in a dialogue.

11. Giorgio Milani's words
Of all things only names remain, and of all possible signs
the alphabet's letters.

12. Francesco Bonami's words
The art: a real need of society and individuals that reduces
the concrete needs and trigger those of the imagination.

13-14. *Four Quartets* **by T. S. Eliot**
Footfalls echo in the memory
Down the passage which we did not take
Towards the door we never opened
Into the rosegarden.

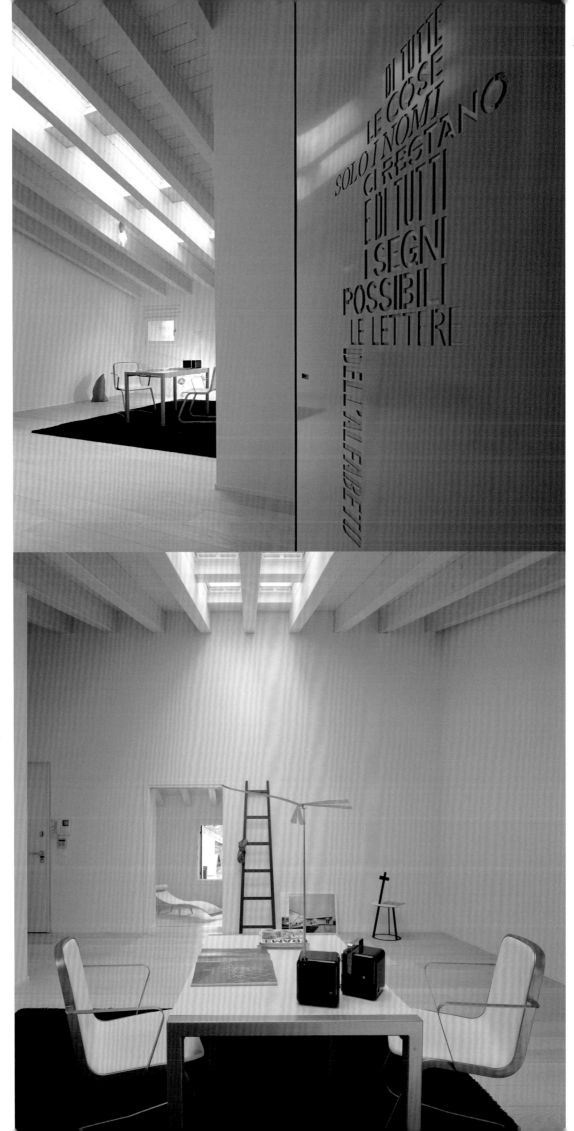

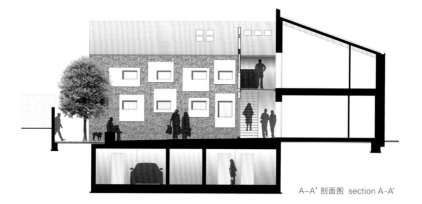

A-A' 剖面图 section A-A'

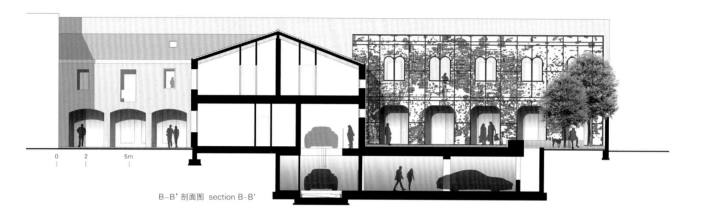

B-B' 剖面图 section B-B'

项目名称：Corten Apartments
地点：Via Veneto 5d, 30030 Vigonovo Venice, Italy
建筑师：3ndy Studio
立面设计艺术家：Giorgio Milani, Philippe Daverio
甲方：Cosmo Immobiliare snc.
总建筑面积：300m²
竣工时间：2011
摄影师：©FG+SG Architectural Photography (except as noted)

塔利亚剧院

Gonçalo Byrne Arquitectos + Barbas Lopes Arquitectos

葡萄牙两代建筑师联手复原了里斯本的这家毗邻城市动物园的新古典主义风格剧院的废墟。建筑的废墟被一层混凝土外壳所覆盖，并且延伸至一座玻璃馆，形成一处可用于举办活动和演出的多功能空间。

塔利亚剧院于1843年竣工，属于Farrobo伯爵的私人财产。剧院坐落于里斯本的郊区，面向一座宫殿和花园，这也是这位贵族的私产。Farrobo伯爵热衷于艺术，塔利亚剧院曾被用于上演舞台剧和歌剧，也曾举行奢华的派对供伯爵娱乐。1862年，一场大火烧毁了塔利亚剧院的镀金木材、镜子和枝形吊灯等豪华装饰。那时Farrobo伯爵已家财散尽，死时甚至身无分文。

150年以来，塔利亚剧院一直是残垣断壁。后来里斯本的郊区向周围下行延伸，城市动物园与其相邻，成为这个残余结构的较为奇异的背景。到2008年，葡萄牙教育和科学部委托发起了一项研究，即如何把塔利亚剧院重建成一处多功能空间。

这个公共机构现在就坐落在宫殿的前面。里斯本当地的Gonçalo Byrne建筑事务所和Barbas Lopes建筑事务所对塔利亚剧院进行了重建。

为了保持塔利亚剧院的旧墙壁原有的样子，其外部全部采用赤褐色的混凝土外壳来进行覆盖，形成一个巨大的整体结构。这个外壳覆盖了原先的观众区和舞台的体量，高度为23m。在内部两处空间里，原来的废墟原样保留下来，使其本身就是一个奇观。采用最小型的技术设施便能创造一个舞台，以满足展览、峰会、音乐会、派对或是播放节目的需要。

毗邻剧院的是一座单层玻璃馆，即一座全玻璃，可承担接待、服务、自助餐厅等附属功能的建筑。这座新翼楼环抱着塔利亚剧院，面向邻近的繁忙大街，其连续的玻璃嵌板构成的表面反射了周围的景物，显得富丽堂皇。入口是利用原始的门厅建成的，为新古典主义风格，包括内部由泡沫聚苯乙烯型材制成的带槽纹的檐壁，而外部则是对前门廊和大理石狮身人面像进行了修复。

主立面的门楣中心采用铜字篆刻的字母构成了喜剧女神塔利亚的箴言，其原文为拉丁文"Hic Mores Hominum Castigantur"，大意是"在这里人类的作为将受到惩罚"，这一刻印再次被放置在此处。里斯本塔利亚剧院的重建将建筑的新旧部分相结合，形成一个城市综合体，可以看见邻近动物园的视野。它使人们重温过去，而这种过去作为一种幻想、想象和市民生活的地方而存在着。

Thalia Theater

Two different generations of Portuguese architects reconverted together the ruins of a neoclassical theater in Lisbon, next to the city zoo. The remains of the building are covered by a shell of concrete and extend onto a glass pavilion, forming a multipurpose space where events and performances can take place.

The Thalia Theater was inaugurated in 1843, as a private venue for the Count of Farrobo. It was located just outside of Lisbon, facing a palace and gardens also owned by this aristocrat. As a lover of the arts, the Count of Farrobo used the Thalia Theater to stage theater and opera shows as well as extravagant parties for his own amusement. In 1862, a fire burned down the building with all its luxurious decoration of gilded wood, mirrors, and chandeliers. By then, the Count of Farrobo had lost his fortune and eventually died penniless.

For almost 150 years, the Thalia Theater remained in ruins. While the outskirts of Lisbon sprawled round its whereabouts, the city zoo was located next to it providing an exotic background to the derelict structure. In 2008, the Portuguese Ministry of Education and Science commissioned a study to reconvert the Thalia Theater into a multipurpose space.

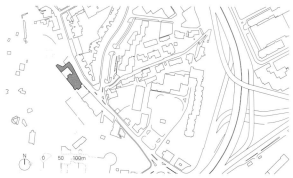

This public institution is presently located in the palace just in front. Gonçalo Byrne Arquitectos and Barbas Lopes Arquitectos, both Lisbon-based, made the project.

In order to retain the old walls of the Thalia Theater, the exterior is entirely covered by a shell of terracotta concrete that forms a massive and monolithic body. It is composed of the original volumes of the audience and the stage, 23m high. Inside these two voided spaces, the ruins are left untouched in a spectacle of their own. Minimum technical fixtures create an arena that can be adjusted to several uses such as exhibitions, summits, concerts, parties or broadcasts.

An adjacent single-story structure, entirely glazed, houses additional program such as a reception, services, and a cafeteria. The new wing embraces the Thalia Theater and faces the busy street next to it with a continuous surface of glass panes that mirror the environs with gilded reflections. The entrance is made by the original foyer, reconstructed in a "neoclassical" style including a fluted frieze on the inside built with styrofoam profiles. On the outside, the front portico and marble sphinxes were restored.

Bronze letters at the tympanum of the main facade spell out the motto of Thalia, the muse of comedy. The original Latin inscription, "Hic Mores Hominum Castigantur", was placed once again. In other words, "Here the Deeds of Men Shall Be Punished". The project for the reconversion of the Thalia Theater in Lisbon combines the old and new parts of the building into an urban ensemble with views to the nearby zoo. It brings back the presence of the past as a place for fantasy, imagination, and civic life.

Gonçalo Byrne Arquitectos + Barbas Lopes Arquitectos

项目名称：Thalia Theater
地点：Lisbon, Portugal
建筑师：Gonçalo Byrne Arquitectos+Barbas Lopes Arquitectos
合作商：Hugo Ferreira, João Neves, Jan Vojtíšek, Lígia Ribeiro, Luca Martinucci, Tânia Roque
工程师：AFAconsult, Natural Works
承包商：ACF
甲方：Portuguese Ministry of Education and Science
用地面积：3,500m²
总建筑面积：1,150m²
造价：EUR 2,700,000
竣工时间：2012
摄影师：©DMF (courtesy of the architect)

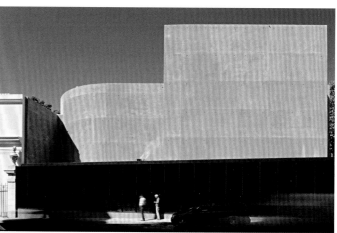

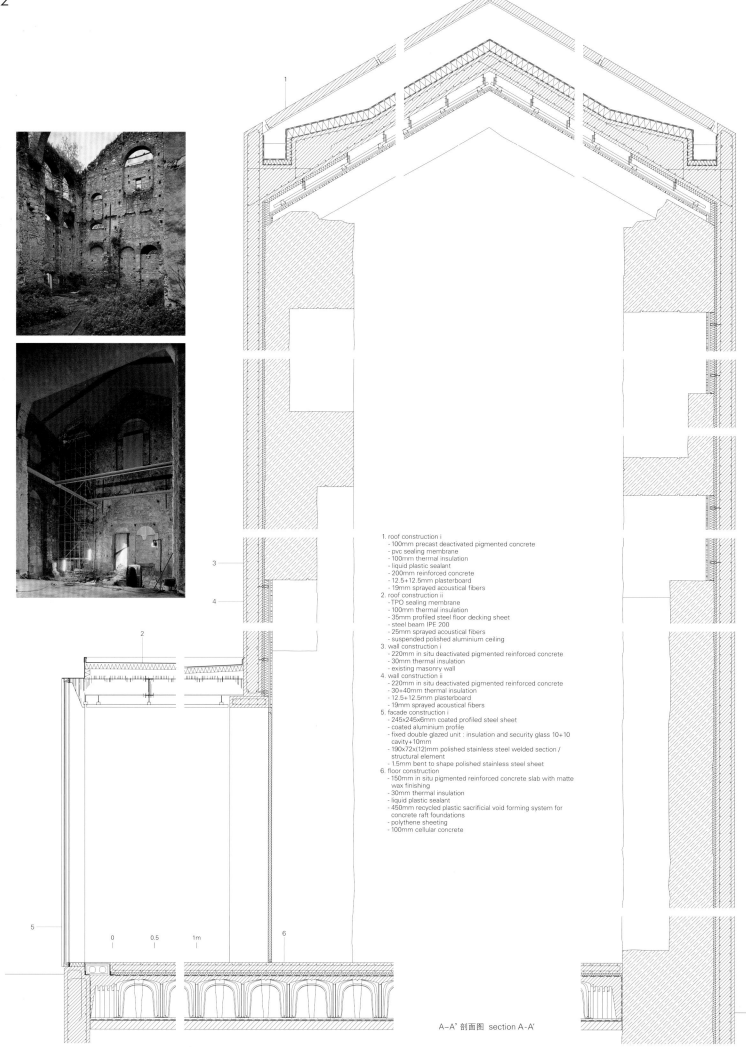

1. roof construction i
   - 100mm precast deactivated pigmented concrete
   - pvc sealing membrane
   - 100mm thermal insulation
   - liquid plastic sealant
   - 200mm reinforced concrete
   - 12.5+12.5mm plasterboard
   - 19mm sprayed acoustical fibers
2. roof construction ii
   - TPO sealing membrane
   - 100mm thermal insulation
   - 35mm profiled steel floor decking sheet
   - steel beam IPE 200
   - 25mm sprayed acoustical fibers
   - suspended polished aluminium ceiling
3. wall construction i
   - 220mm in situ deactivated pigmented reinforced concrete
   - 30mm thermal insulation
   - existing masonry wall
4. wall construction ii
   - 220mm in situ deactivated pigmented reinforced concrete
   - 30+40mm thermal insulation
   - 12.5+12.5mm plasterboard
   - 19mm sprayed acoustical fibers
5. facade construction i
   - 245x245x6mm coated profiled steel sheet
   - coated aluminium profile
   - fixed double glazed unit : insulation and security glass 10+10 cavity+10mm
   - 190x72x(12)mm polished stainless steel welded section / structural element
   - 1.5mm bent to shape polished stainless steel sheet
6. floor construction
   - 150mm in situ pigmented reinforced concrete slab with matte wax finishing
   - 30mm thermal insulation
   - liquid plastic sealant
   - 450mm recycled plastic sacrificial void forming system for concrete raft foundations
   - polythene sheeting
   - 100mm cellular concrete

A-A' 剖面图 section A-A'

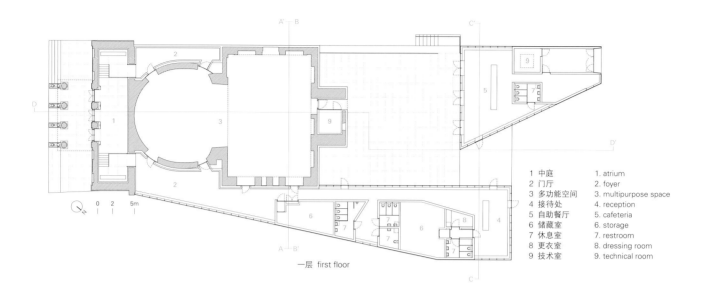

1 中庭	1. atrium
2 门厅	2. foyer
3 多功能空间	3. multipurpose space
4 接待处	4. reception
5 自助餐厅	5. cafeteria
6 储藏室	6. storage
7 休息室	7. restroom
8 更衣室	8. dressing room
9 技术室	9. technical room

一层 first floor

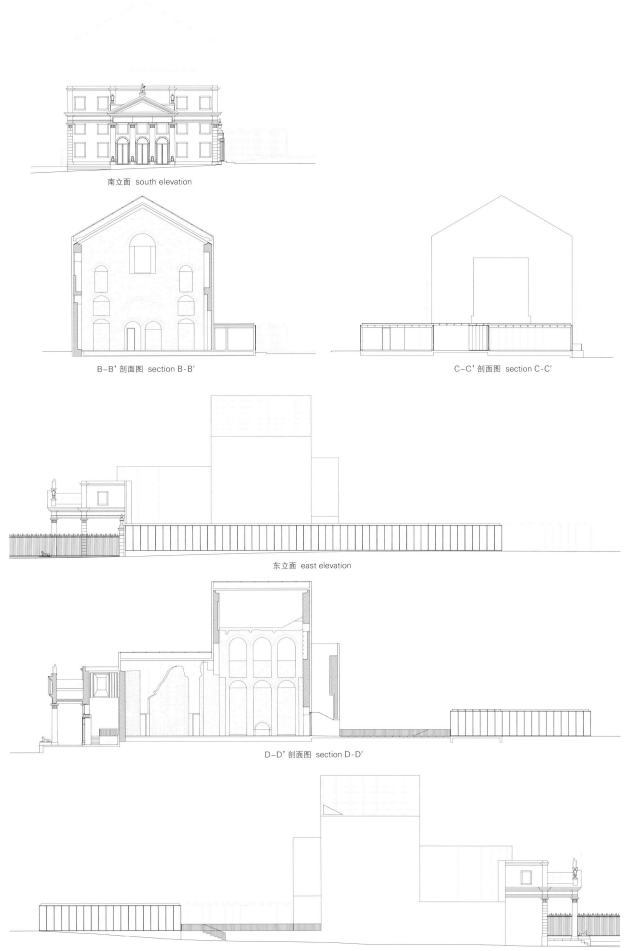

南立面 south elevation

B-B' 剖面图 section B-B'

C-C' 剖面图 section C-C'

东立面 east elevation

D-D' 剖面图 section D-D'

西立面 west elevation

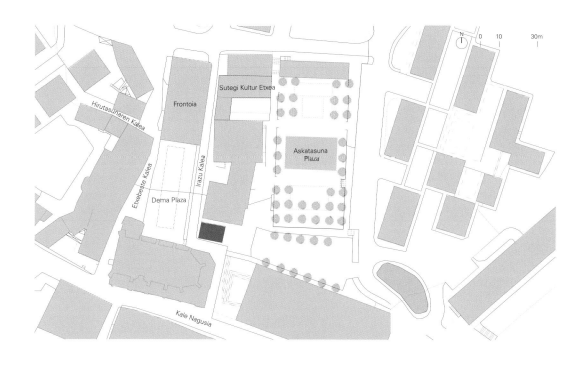

Potxonea住宅

OS3 Arkitektura

Potxonea住宅是一个街区的最后一座建筑，这个街区是市中心的脊柱。这座住宅是一个有历史背景的超凡建筑，毗邻的建筑划分出两处大型公众区域：Dema广场和最近建成的Askatasuna广场。

两个广场的不同反映在该建筑物的立面上，因此产生了两种完全相反的风格，面对Dema广场的立面是由承重墙、小型洞口和方石饰面组成；而面向Askatasuna广场的则是更加朴实的立面，它由带有大型洞口的木质桁架组成。在这种环境内，Potxonea住宅建筑是Irazu街的最后一幢建筑，它和周围的建筑物由一条狭窄的1.7m宽的防火墙通道隔开，这条通道围以栅栏，且从未被使用。

这座建筑的原有平面很小，空空荡荡而且结构破旧。城市保护特别计划将场地划入基本保护级别，这意味着要保持建筑物的立面一直面向Dema广场。

为了最大化地利用规划的区域，建筑师提出了一个加以控制的扩张提案：1.通道只占用地面层；2.根据民事法典，该建筑物应和周围的建筑物相隔3m，所以该建筑只向Askatasuna广场略微扩充一些面积，且维持形成Irazu大街的所有建筑的后退距离；3.建地下室，以增加可使用的楼层空间；4.尽可能地利用特别计划允许的探出部分：50%的南立面都探出了30cm。扩建部分的面积有所增大，因此原有的立面和探出部分之间的过渡空间被保留下来。

在内部，纵向的交流元素设置在北立面中，使这些区域具有更好的照明和视野。通过使自然光线最大限度地遍布这个区域，室内空间的视觉和空间交流也强化了人们对空间的感知力。

主要入口面向Askatasuna广场，在这个街区是独一无二的，能够产生突出入口的更加舒适的空间。

对于建筑新表皮的处理方式可由以下几种因素来定义：

与原承重墙的小孔相比，新孔的规格更大一些，这样更多的自然光能够进入室内，感觉更加明亮。

同时，在现存的文化建筑中，从外面可以看到室内的活动是非常有趣的。墙孔的分布是非对称的，同时也没有呈现垂直的布局，突出了新体量的轻巧性。

建筑的西立面恢复了其原来的带屋檐的三角形屋顶，新体量从原有的平面中凸出来，其覆盖的新表皮与屋檐融为一体。然而在入口处，木材保持其天然的饰面，这种木饰面延续至建筑的内部，并且贯穿了整个小巷的长度。通过采用如此简单的设计和镶嵌的细木工木门，一个带顶的入口区从立面的其他洞口中凸显出来。

总之，在遵守法律的前提下对各种可能性进行研究，建筑师设计了多面的建筑作品，古老与新潮的交相辉映使得细节更丰富。

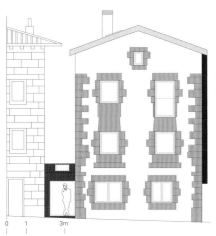

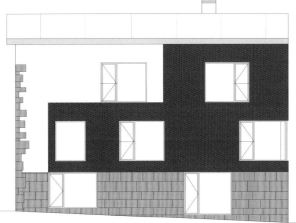

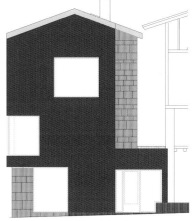

西立面 west elevation   南立面 south elevation   东立面 east elevation

Casa Potxonea

The Potxonea is the last of a block that provides the backbone to the downtown, being a remarkable building with a historical background. The adjoining buildings separate two large public spaces: the Dema Square and Askatasuna Square, more recently built.

The difference between these two squares is reflected in the building facades, resulting in two totally antagonistic styles. The one overlooking Dema Square is composed of load-bearing walls, openings of small size and ashlars stone finishes, while the front facing the Askatasuna Square shows more modest facades consisting of timber trusses with larger openings. Within this context, the Potxoenea building was the last in Irazu Street and was separated from the adjoining building by a narrow 1.7m firewall alley that was fenced and unused.

This building had a very small plan, empty and in a very poor structural state. The Downtown Protection Special Plan gave this site a basic protection level which meant keeping the facade facing Dema Square.

In order to maximize the quality of the areas included in the program, a controlled expansion was proposed: 1. The alley was occupied on the ground floor only; 2. A small extension was raised to Askatasuna Square, respecting 3m separation from the adjacent building, pursuant to Civil Code and maintaining the setback schemes of the buildings that make up Irazu street; 3. Constructing a basement for more usable floor space; 4. Exhausting projections permitted by the special plan: 30cm in 50% of the south facade. This extension was raised so that a transitional space was left between the kept facade and projection.

Inside, the vertical communication elements were placed in the north facade, thus liberating areas with better lighting and views. The quality of the interior spaces with visual/spatial communication by making natural light reach the maximum area also expands the perception of the spaces.

The main entrance is from Askatasuna Square, unique in this block. It allows generating a more suitable space that emphasizes the entrance.

Treatment of the new skin is defined by the following factors:
Against the small holes of the remaining bearing wall, the new holes are broader, achieving more efficient natural lighting and a feeling of lightness.

At the same time, and in the case of a living cultural building, it's interesting that what is happening inside can be seen from outside. The distribution of holes not aligned vertically reinforces the weightlessness of the new volume.

The west facade regains its original gable roof with eaves configuration, while the new volume protrudes from the previous plan absorbing the eaves under the same cover. In the entrance area, however, the wood is left with its natural finish. This continues inside the building throughout the entire length of the alley. With this simple resort, and imbedding the joinery of the door, a covered access area that stands out from the rest of openings in the facade composition is achieved.

In conclusion, by a real study of the possibilities left after the application of the laws, a multiple faceted building is created. The richness in details comes from the interaction of the old with the new.

OS3 Arkitektura

项目名称：Casa Potxonea
地点：Irazu Kale, 4, Usurbil
建筑师：Ainara Sagarna, Maialen Sagarna, Juan Pedro Otaduy
投资商：City of Usurbil
用地面积：139.18m²
使用面积：382.76m²
造价：EUR 841,128.17
设计时间：2008
竣工时间：2011
摄影师：courtesy of the architect-p.47[bottom]
©Jorge Allende(courtesy of the architect)-p.47[top], p.50, p.51
©Jorge Allende-p.46, p.48, p.49

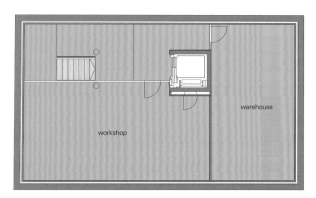

地下一层 first floor below ground

一层 first floor

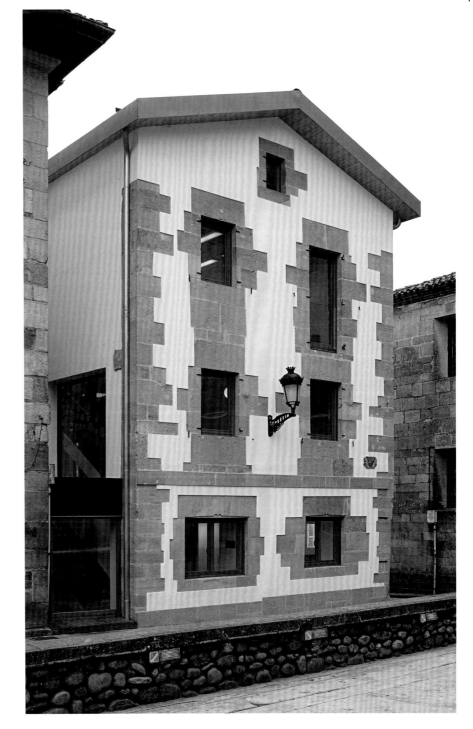

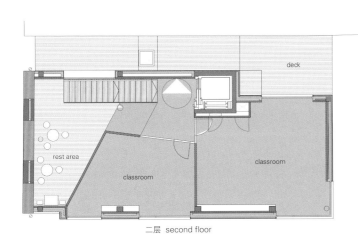

二层 second floor

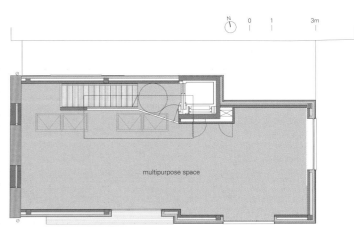

三层 third floor

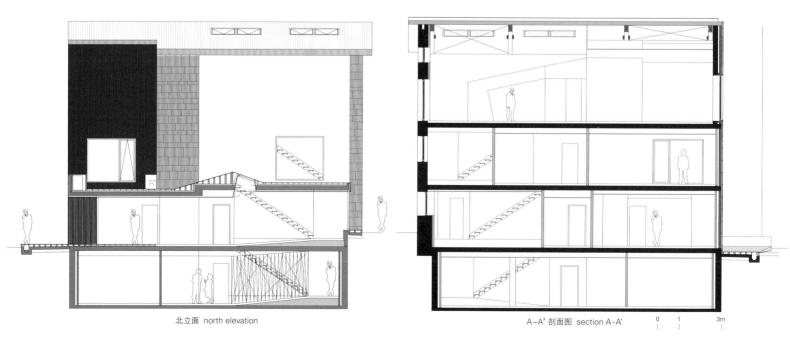

北立面 north elevation

A-A' 剖面图 section A-A'

1. ainc top
2. ainc gutter
3. EGO_CLT-160 panel
4. empty batten
5. 28x38mm batten
6. Imerys OMEGA 10 tile
7. douglas fir GL-24h
8. termoarcilla serie 24 tie bar
9. painted pointing
10. double hollow bricks
11. wall cavity
12. project polyurethane
13. water repellent pointing
14. termoarcilla 300x240x190mm
15. plaster
16. DM wooden skirting board
17. Mastertop 1225. epoxi mortar
18. polypropylene fibre reinforced mortar
19. extruded polyethylene
20. polyethylene steam barrer
21. zinc piece
22. polydros fiberglass thermal insulation
23. black painted batten
24. 26x90mm red cedar wooden slat
25. reinforced concrete lintel
26. zinc raindrop
27. delta facade sheet
28. wooden counterframe
29. water repellent pointing
30. pladur roof
31. stadip 3+3+3 glass railing
32. 5/12/6 double glass
33. meranti wooden frame
34. natural sandstone
35. brick
36. waterproof sheet
37. reinforced concrete wall
38. edgewise beam
39. Feltemper 300p antipunch sheet
40. Fiiltron slab
41. Rhenofol cg sheet
42. wall
43. reinforced concrete slab
44. extrude polyethylene
45. air circulation pipe
46. mortar canal
47. natural ground
48. gravel
49. PVC sheet
50. neoprene connection
51. concrete foundation
52. reinforced concrete girder
53. draining pipe

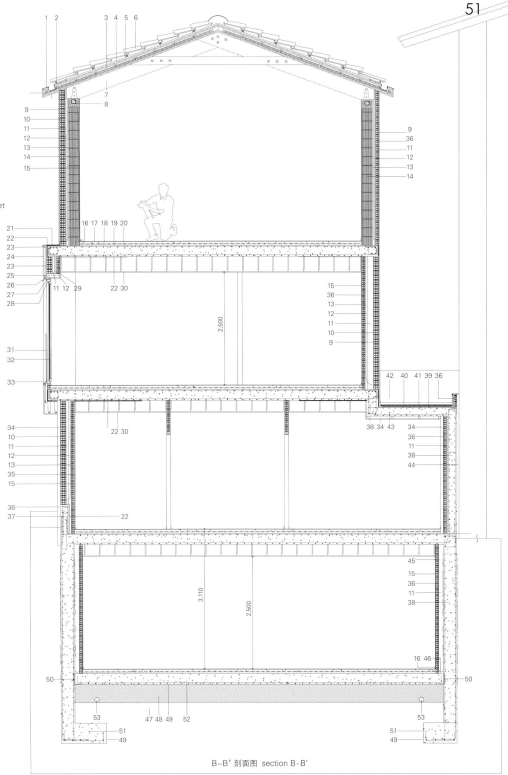

B-B' 剖面图 section B-B'

锚固与飞翔
——挑出的住居
Podia, Plinths and Flying House

可追溯到古典建筑时期甚或更早时候的主要构造原理之———三分法是将每座建筑都分为底座、建筑主体和屋顶。纵观历史,每一部分均与城市和居住者有着特殊的关系:底座和街道及路人直接相连;主体部分是构成城市的"经络",以三维立体的形式存在;屋顶决定建筑的高度,标志着建筑的竣工。根据建筑规模的大小,这三部分结合在一起发挥着不同的作用,使原本孤立的建筑瞬时变成了一个综合体。试问如果其中一个关键部分不复存在会如何呢?那么这座建筑是否依然会呈现复杂的外观并且发挥多种功能呢?另外,这座建筑还能算得上是综合体吗?抑或根本就不完整?内部的结构会受到影响吗?换言之,如果建筑物不是根深蒂固地坐落在地基上而是设计成悬挑式的,那结果会如何?如果在不平坦的地面上建造的建筑物呈现凌空架势会如何呢?

One of the main compositional principles since the time of Classical architecture, and even before, has been the tripartition of every building into plinth, body and roof. Each part, throughout history, has had a specific relationship with the city and its inhabitants: the plinth fosters direct connections with the streets and passersby, while the body constructs the urban tissue of the city in its tridimensional extent, and the roof determines its end in height, concluding the building. The overall combination plays various roles as the scale changes, turning a single object (the building) into a complex element. But what happens if one of these crucial compositional elements goes missing? Can the building still embody complexity and perform its various activities? Moreover, is it still a complex or is it now incomplete? What happens to its internal organization? In other words, what if a building is not solidly rooted in its terrain but is conceived as being perched? What if the building takes flight over uneven terrain?

坦格尔伍德2号住宅 / Schwartz/Silver Architects
Algarrobos住宅 / Daniel Moreno Flores + José María Sáez
BF住宅 / OAB – Office of Architecture in Barcelona + ADI Arquitectura
X住宅 / Cadaval & Solà-Morales
贝兰达住宅 / Schmidt Arquitectos
素风宅 / acaa/Kazuhiko Kishimoto
Hanare住宅 / Schemata Architects
纳克索斯岛避暑别墅 / Ioannis Baltogiannis + Phoebe Giannisi + Zissis Kotionis + Katerina Kritou + Nikolaos Platsas
锚固与飞翔——挑出的住居 / Silvio Carta

Tanglewood House 2 / Schwartz/Silver Architects
Algarrobos House / Daniel Moreno Flores + José María Sáez
BF House / OAB – Office of Architecture in Barcelona + ADI Arquitectura
X House / Cadaval & Solà-Morales
House in Beranda / Schmidt Arquitectos
Wind-dyed House / acaa/Kazuhiko Kishimoto
Hanare House / Schemata Architects
Summer House in Naxos / Ioannis Baltogiannis + Phoebe Giannisi + Zissis Kotionis + Katerina Kritou + Nikolaos Platsas
Podia, Plinths and Flying House / Silvio Carta

米开朗基罗的大卫雕塑，1500—1504年
Michelangelo's David, 1500~1504

印度克久拉霍拉希玛纳神庙，约公元930—950年
Lakshmana Temple in Khajuraho, India, CIRCA 930~950 A.D.

墩座

墩座作为核心元素在古典建筑中被广泛采用。直白地讲，墩座即平台，在此之上矗立一个重要物体，可能是建筑、雕塑或代表胜利的立柱。平台的作用是提升建筑物的离地距离，更好地利用周围环境。为了充分发挥墩座的价值，有一点需要注意，墩座上的物体必须具有一定重要性，例如纪念碑、公共建筑、著名人物或记载某一事件的雕塑。Rosalind Krauss在1979年发表的一篇文章[1]里对此进行了详细阐述，她在文章中试图描述20世纪70年代末期雕塑和艺术作品经历了怎样的变化，虽然主要关注点在雕塑建筑方面。鉴于雕塑"一般都富有象征意义并且是垂直的"，它们的基架"是整个结构中的一个重要部分，因为它起到协调场地和象征性标志的作用"。[2]

从不同的角度观察墩座，不难发现在墩座上竖立物体（尤其是建筑），建筑方法和结构布局自有史以来已经发生了诸多变化。尽管变化之多不胜枚举，我们还是可以通过简单地回顾几个重要例子把从最初到目前所经历的演变过程有机地联系在一起。在古典建筑中，墩座是建造寺庙、长方形会堂和宫殿的重要组成部分。它不仅是建造寺庙之初就需要的平台，能提高稳固性及在周边环境中的可视性，同时也被认为是建筑本身不可或缺的一部分。柱上楣构、立柱以及墩座或者叫底座是寺庙垂直分隔结构的三大主要元素，墩座或底座辅以其他部分能够发挥出寺庙独特的结构和象征意义，与整座建筑浑然一体。

底座随着时间的流逝也根据文化的演变发生了翻天覆地的变化。一方面，底座并不一直都是独立的元素。例如在建造金字塔时，底座是镶嵌在整座"建筑"里面的，而不是像希腊或罗马建筑那样独立于建筑物之外。位于韩国西部益山市的弥勒寺宝塔的底座比宝塔本身还大，底座在地面高度上设有多处入口。与之相反，印度克久拉霍神庙的墩座无论是面积还是高度都非常突出，把整个建筑群的各个组成部分紧密联系起来。不难想象，在当今建筑中依然随处可见采用底座的做法，只是根据时代、地理位置和文化的不同，底座对于整座建筑

The Podium

Classical architecture evinces widespread use of the podium as a crucial element. Bluntly, a podium is a platform on which there sits a significant object, whether a building, a sculpture or even a triumphal column. The platform is meant to raise the object a short distance from the terrain, so that it imposes over its surroundings. It is significant to note that, in order to merit a podium, the object must be of a certain importance or significance, such as a monument, a public building, or a statue of a stand-out person or episode. Though focusing primarily on sculptural production, Rosalind Krauss wrote a clarifying article[1] in 1979, in which she tries to describe how statues and artistic production were changing in the late seventies. Given that sculptures are "normally figurative and vertical", their pedestals *"(are) an important part of the structure since they mediate between actual site and representational sign."*[2]

Observing the podium from a quantitative angle, one may notice that the operation of staging an object (in particular a building) has historically undergone many different means and configurations. Although the list of these could potentially be limitless, we can briefly recall a few significant examples to help us trace a sort of continuity from the beginnings up to the present. In Classic architecture the podium was a fundamental part of the composition of temples, basilicas and palaces. It was not only the preliminary platform on which a temple should be erected, providing stability and more visibility from the surroundings, but was considered a constitutive element of the object itself. Along with the entablature and the columns, the podium or plinth was one of the three main elements of the vertical partition of the temple, and – as with any other part – it had a specific constructive and symbolic meaning, in harmony with the whole.

The plinth has varied significantly by culture over time. For one thing, it has not always been an independent element. In the pyramids, for instance, the plinth is embedded into the entire main shape of the "building" and is not considered separate, as in Greek or Roman architecture. The pagoda of the MiReukSa Temple in IkSan (in western Korea) presents an enlarged plinth containing the entrances at the ground level. Conversely, India's Khajuraho Temple presents an extended (in area and height) podium which ties together the various emerging parts of the complex. As is easy to imagine, the plinth remains present in architectural compositions practically everywhere, although it alters its meaning, shape or importance for the entire building depending on the

圆形别墅，意大利威尼托维琴察省，安德烈亚·帕拉迪奥设计，1567—1570年
Villa Rotonda by Andrea Palladio in Vicenza Veneto, Italy, 1567~1570

所代表的意义、形状及重要性有所不同。我们有必要关注一下底座在伊斯兰建筑[3]中的处理方式。伊斯兰的建筑风格独特，例如在伊朗伊斯法罕的国王清真寺或者突尼斯的凯万大清真寺等十分独特的项目中，最显著的特点是整个建筑群的底座下面是全空的，[4]但是空间布局借助拱廊和柱廊显得丰富多样。[5]

如果我们跳转到意大利文艺复兴时期的宫殿，会发现底座在意式宫殿中起着装饰的作用。似乎底座已经失去了对于寺庙的结构意义，设计更加注重整座建筑的各个组成部分，包括建筑设计蓝图本身，后期逐渐发展到注重更大范围的城市布局。在法尔内塞宫这样的宫殿中，建筑下部分成三部分，突出展现与街道相连的部分，如此一来，"宏伟的底部"显得与众不同，高出周围一截而又不至于成为公共活动区域。如果市区外风景抢眼，底座就会再次出现，作为一个抬高的平台，使整座建筑与周围的环境相比更占上风。帕拉迪奥设计的圆形别墅便是这种布局的绝佳范例。

墩座的消失

纵观整个建筑发展史，不难发现，人们曾尝试不用底座打造建筑，或者从更广泛的意义上来说，利用完全不同的建筑方式。20世纪勒·柯布西耶在他的"新建筑的五个特点"中介绍了他在这方面的努力，他的建筑理念是采用底层架空柱来提升建筑离地距离，他认为，"房屋离地，可以远离潮湿和黑暗，屋底下面也可以建花园。"[6]萨伏伊别墅建筑本身的比例体现了上述原则，马赛公寓可谓该原则的城市版，在这个项目中，"公共地毯"即绿地草坪（译者注）肆意流动在城市的地平面上。

然而到了后现代，建筑师们借鉴古典建筑中的象征手法，再一次把墩座引入建筑当中（例如查尔斯·摩尔在新奥尔良设计的意大利广场），广场的建成再次让人们开始自由采用传统的建筑元素（至少在想法上如此）。基于这种自由，出现了埃森曼设计的住宅或卢西奥·帕萨雷利在罗马设计的混合用途结构（这两个项目的底座在多个楼层上

period, geographic area and culture. An important observation should be made with regard to the treatment of the plinth in Islamic architecture[3]. Visible in quite distinct projects, such as the Shah Mosque in the city of Isfahan (Iran) or the Great Mosque of Kairouan (Tunisia), is a significant emptying from the complex of the plinth,[4] along with the appearance of quite rich spaces characterized by arcades and porticos.[5]

If we jump to the palaces of the Italian Renaissance, we see the plinth assuming the connotation of a decoration. It seems to lose its structural meaning (which it had for the temples), while emphasis is given to its part in the overall composition, both of the "drawing" of the building per se, and, later, in the more extended urban composition. In palaces like the Palazzo Farnese the lower part announces a tripartite rhythm and openly declares the part of the building which relates more to the street level, differentiating the "noble floors", elevated and therefore separated from public activities. Outside the city, when the landscape component becomes predominant, the plinth reappears as an elevated platform in order to establish a more privileged relationship of the building with its surroundings. The Villa Rotonda designed by Palladio is a clear example of such an arrangement.

Disappearance of the Podium

Throughout the history of architecture one may observe several attempts at building without plinths or, more generally, vastly different approaches. Le Corbusier discusses such efforts in his "Les Cinq Points d'une Architecture Nouvelle" in the twenties, describing his principle of employing pilotis to elevate the mass off the ground: *"La maison est en l'air, loin du sol humide et obscur, le jardin passe sous la maison. (The house is in the air, away from the wet and dark, and the garden runs under the house.)"*[6] The consequences of this idea can be seen at the scale of the building itself in the Ville Savoie, but also in their urban implications in Marseille's Unité d'Habitation, where the "public carpet" is, free to flow into all the urban surface at the ground level.

While the Postmodern – by employing figurative elements from Classical architecture – brings the classical podium to buildings once again (think of Charles Willard Moore's Piazza d'Italia in New Orleans), its conclusion has left complete freedom (at least in thinking) about traditional elements of building. From this freedom have arisen such buildings as Eisenman's Houses or Lucio Passarelli's mixed-use structure in Rome (where the plinth houses various programs on several floors) up to such extreme forms as

蛇形展馆，2009年，SANAA事务所设计，立面拔地而起，没有任何中断的迹象
Serpentine Gallery Pavilion 2009 by SANAA, the facade rises from the ground without announcing any breaking point

容纳了不同的功能空间），后来建筑形式发展到了极致，如位于毕尔巴鄂由扎哈·哈迪德或盖里设计的古根海姆博物馆，"基座"成为建筑的一个无足轻重的元素。然而纵观建筑发展历史，我们还可以看到早在现代主义阶段就开始的另一个平行发展的趋势：完全忽视墩座的作用。从这个角度来看，我们可以列出一长串建筑物的名字，例如诺伊特拉设计的别墅或从时间上来讲离我们更近的日本SANAA事务所设计的几个项目，这些建筑的立面拔地而起没有任何中断的迹象。从这个层面上来讲，这类建筑从外面看起来仿佛每一层的平面设计都是一样的，而忽视了建筑垂直方向上"起点"和"终点"的存在。如此一来，从理论上讲这类建筑可以无限高或者建筑高度可以被复制。

栖息

正如上文所言，底座的存在、结构及重要性在建筑史上发生过重大变化。底座可以是中空的抑或是切开的空间。底座可以是独立的建筑元素、结构支撑部分和重要的建筑组成部分，它们也可以被完全忽视，和整座建筑毫无关联。从下面介绍的项目中可以看出现行的做法——建筑没有基座自成一体，使住宅悬挑于高处。

这里展示的项目依据建筑与选址地形的关系主要分为两大类。

第一类住宅一侧固定，另一侧则是悬于美丽的景色之上。巴塞罗那OAB建筑事务所和ADI建筑师事务所在西班牙卡斯特利翁省的一个名叫Borriol的小镇设计的BF住宅呈现独立的整体，可以看出由两个鲜明的体量组成，其中一个从路边的入口处开始便扎根在地底，另一个悬挑着，俯瞰全景，内有厨房、客厅和主卧。这两个体量中间有一个中空空间，而且两部分设计一致。房屋结构布局独特，营造出不同的景致，建筑利用自然的地形，搭配周围的自然景观，处处洋溢着和谐。从住宅的前面、下面和上面都可以观赏到风景。

BF住宅以纵轴为依托，Schwartz/Silver建筑师事务所主持设计的坦格尔伍德2号住宅恰恰相反。住宅位于美国马塞诸塞州的西斯托克布里奇，以伯克希尔山全景为背景，垂直于背景建造，从中间开始

Zaha Hadid's or Gehry's Guggenheim in Bilbao, where the "base" of the building becomes an insignificant element. However, as a general overview, one may recall a parallel trend which began even in the Modern period: a total indifference to the podium. From this perspective one can list such buildings as Neutra's villas or – closer to our time – several of SANAA's projects, where the facades simply rise from the ground without announcing any breaking point. In this sense, such buildings appear from outside as vertical extrusions of a generic floor plan, neglecting the existence of the "start" and "end" point of the vertical progression of the building. Such buildings could theoretically be endless or genetically repeated in height.

Perched

As mentioned above, the plinth's presence, configuration and importance have varied quite significantly throughout the history of architecture. Plinths can be voids or carved spaces. They can represent a clear separation of elements, structural support, a significant part of the building – or they can simply be ignored, as an irrelevant presence in the overall composition. The projects that follow are an attempt to trace a possible current condition in which the base of the building is physically missing as an autonomous compositional element, leaving the house perched.

The projects here presented can be divided in two main groups, depending on the relationship they establish with the terrain on which they sit.

The first group encompasses those houses that seem to be hinged on one side of the volume, leaving the other part overhanging outstanding views. The BF House, designed by OAB – Office of Architecture in Barcelona + ADI Arquitectura in Borriol, Castellón de la Plana, Spain for instance, presents a monolithic block in which two distinct volumes can be observed, one rooted in the ground at the entrance from the road, and the second cantilevered over the panorama, housing the kitchen, living room and master bedroom. The two volumes are rendered homogeneous by means of an inner central void. The particular configuration of this house allows for a variety of sights in which the natural terrain of the plot is in dialogue with the natural view of the surroundings. Views emerge through the front part of the house, but also underneath and above it.

While the BF House is hinged on the longitudinal axis, the Tanglewood House 2 by Schwartz/Silver Architects offers an example of the opposite principle. This house is built along a line perpendicu-

BF住宅，整座建筑的一部分悬挑在全景上方
BF House, a monolithic block with the part cantilevered over the panorama

Algarrobos住宅，大型悬挑延伸到峡谷上方
Algarrobos House with the large cantilevers, stretched out to the ravine

略微倾斜。因此住宅通过混凝土地下室锚固在地面上，正如建筑师所言，住宅从正面"突然崛起"。住宅底座的前半部分依附地形得以固定，后半部分偏离原来的走势，建在挖掘出的圆形平面地基上。从某种程度上讲，墩座在这里沿线性走势发挥双重作用：作为基础承载整座住宅，另外一半隐藏，使部分住宅像是在飞翔。

Cadaval & Solà-Morales建筑事务所设计的X住宅集BF住宅的纵长和坦格尔伍德2号住宅的横长特点于一身。X住宅位于西班牙巴萨罗那的卡布瑞斯，角度稍偏，从正面可以看到很多不同的小平面。住宅的墩座起到牢固稳定建筑主体后方的作用，但是从前面又看不出来，恰好给面向全景的建筑立面下方的一个陡坡腾出空间。

和BF住宅相似，由José María Sàez和Daniel Moreno Flores设计的Algarrobos住宅叠立在厄瓜多尔的Puembo，利用底座结构呈现出建筑复杂但不失庄重的感觉，同时也能观赏到多种多样的风景。建筑利用了钢结构，实际上跨度变得更自由，设置空间和空隙的自由也更多。建筑只用部分墩座的做法被认为不牢固，但是空间大了，视野更开阔，行动更方便。

第二类建筑主要是那些底座埋入地下的项目。由智利扎帕拉的Schmidt建筑事务所设计的贝兰达住宅特色鲜明，地下室深入地下，垂直设置了多个楼层。住宅有多个组成部分，相互流动穿插，彼此相连，制造出体量外观大小不一、室内布局各不相同的效果。建筑采用推拉和可移动的处理方式营造出多面的整体结构，与倾斜的地势形成的等高线相映成趣。虽然贝兰达住宅朝海滨的立面有一适中的悬挑设计，但是出彩的地方还是底座上呈现的复杂线条，与建筑其他立面紧密相连。在这里，底座不是孤立的建筑元素，而是整个结构的有机组成部分。

我们可以用同样的视角欣赏位于日本神奈川县的素风宅，设计师是acaa建筑事务所的岸本和彦。住宅基础完全埋入地下，但是正面几处开口使内部空间变得独特。底座在整座建筑中不明显，因为住宅依

lar to the panorama of the Berkshire Hills, in West Stockbridge, MA, USA, and is tilted at mid-length. The house is therefore anchored in the ground by its concrete basement and "dramatically rises", as the architects explain, over the front side. The plinth of this house adheres to the terrain in the first half, where it is anchored, and diverges from it in its second half, staging from a circular groundwork which has been carved out and flattened. To a certain extent, the podium here performs a twofold function along a linear development: it roots the house in the ground and disappears in the second half of the length, resulting in one part of the house taking flight.

The longitudinality of the BF House and the transversality of the Tanglewood House 2 somehow appear in combination in the X House by Cadaval & Solà-Morales. This house in Cabrils, Barcelona, Spain, offers multiple facets exposed to the front view at inclined angles. The podium of this house offers a solid ground for anchoring the back of the main volume, while dramatically disappearing in the front, giving space to a steep slope beneath the facade facing the panorama.

Similarly to the BF House, the Algarrobos House by José María Sáez and Daniel Moreno Flores in Puembo, Ecuador, has a plinth which presents a formal complexity allowing for multiple and diverse views. The metal system employed for the construction, in fact, offers large free spans and a larger freedom of space and voids. This sort of fragmented podium for the building is not considered a solid element, but a spatial condition allowing for the permeability of sights and movements.

The second group consists of those projects whose plinth is sunk in the terrain. The House in Beranda by Schmidt Arquitectos in Zapallar, Chile, is characterized by a basement anchored to the ground and layered vertically. The various sections of the building slide into each other to create different conditions in terms of both volumetric appearance and the conditions inside the house. The sliding and shifting operations create a multi-faceted overall volume which plays with the contour lines of the sloping terrain. Although House in Beranda presents a modest overhang for some of the front facade facing the seashore, its main feature is the complexity of the lines of its plinth, directly related to the rest of the building's elevations. Again, the plinth is not a separate element in composition, but appears as an integrated component of the overall volume.

A similar lens can be used to view the Wind-dyed House in Yoko-

素风宅，埋入地下，朝向大海的一侧设有一些洞口
Wind-dyed House, buried in the ground with some openings towards the sea

贝兰达住宅，地下室锚固在地下，垂直设置了多个楼层
House in Beranda with the basement anchored to the ground and layered vertically

地势而建，同时在全景中营造出别有洞天的一番景致。

位于日本千叶，由Schemata建筑师事务所设计的适于隐居的Hanare住宅是我们谈论的话题中的特殊一例，它的地基尤其深入地下。住宅隐于一山峰中，山峰还插入到了建筑体量内。建筑师们建造"山中城堡"的想法通过在半山腰建成悬浮的平房得以实现。悬浮的感觉是借助中空的底座产生的，但看上去像一系列结构支柱，给住宅提供了水平的平台。多亏了"缓冲底座"，房子得以飘浮在山顶上。

关于底座和住宅的话题，希腊纳克索斯岛上的避暑别墅是我们观察到的极端例子。事实上，在这个项目中不太能看清楚底座是否作为独立的一部分和整栋建筑真正分开了，但是可以断言的是，整座别墅本身即是一个底座。这座非同寻常的建筑依斜坡地势建造，避免建筑体量中出现任何斜度或者整个轮廓中出现新的线条。避暑别墅尝试充分利用斜坡地址，房屋和地势合二为一，尽管边缘的直线清楚地表明建筑师想在岩石和高低不平的地势上建造纯粹的建筑体量。屋内的一切都保持了同样的线性效果，包括平屋顶、开放式泳池和连接各部分的台阶。

底座在整个结构中千变万化，不论有无、突出或隐藏，它在这个项目中的设计已经达到了极致并完成了其最终的转变：底座即房屋。

1. Rosalind Krauss，"Sculpture in the Expanded Field"，October, Vol.8., 1979, Spring
2. 同上
3. 尽管几乎不太可能将所有这一类的建筑都包含在内，但我们仍然可以从中了解其基本思想，从而确定我们谈论内容的另一个观点。
4. 对于这类建筑，从内部可以感到到是空的，而在外面看，建筑仍然具有堡垒的坚固特点。
5. 拱廊在建筑史上一直以不同的方式出现，但是在伊斯兰建筑中这类空间的华美却极其独特。
6. 参见http://www.fondationsuisse.fr/FR/architecture2D.html – accessed January 18th 2013.

suka Kanagawa (Japan) by acaa/Kazuhiko Kishimoto. This house is definitely sunk into the ground, yet the several openings toward the front create special spaces within. The perception of the plinth is here not clear, as the house gently accompanies the sloping terrain while simultaneously providing outstanding views over the panorama.

The Hanare House retreat in Chiba, Japan by Schemata Architects presents to our discourse an extreme case of sinking. The project is immersed in a peak which perforates the volume of the house. The architects' idea of building "a castle on a mountain" is elaborated through a suspended flat volume standing in the middle of the peak. The suspension is here created by the empty volume of the plinth, which is conceived as a series of structural pillars, providing the house a horizontal platform. Thanks to the "buffer plinth", the house can float upon the peak.

The Summer House in Naxos (Greece) represents perhaps an extreme point in our observations concerning plinths and houses. In this project, in fact, it is not quite clear whether a plinth actually exists as an element separate from the volume; it can asserted with equal validity that the entire house is itself a plinth. This radical project directly follows the inclination of the terrain, avoiding any sliding of volumes or new lines in the silhouette. The Summer House attempts a total integration with the slope on which it sits, although the straight lines of its edges clearly announce the pure volumetric intention over the rocky and uneven terrain. Everything in the house happens within the same linearity, including the flat roofs, the open pool and the steps connecting one part to another.

The multitude of guises that the plinth has assumed with regard to the entire composition, including in its absence, negation, emphasis or disappearance, has here reached its logical extreme, with the plinth undergoing the ultimate transformation: the plinth is the house. Silvio Carta

1. Rosalind Krauss, "Sculpture in the Expanded Field", October, Vol.8., 1979, Spring
2. ibidem
3. Although it is nearly impossible to encompass the myriad structures which may bear this label, we may yet refer to a general idea in order to locate another coordinate of our discourse.
4. For such buildings, the emptying is appreciable from the inside spaces, while on the outside they still present the robust characteristics of a fortress.
5. Arcades will appear all over the history of architecture in various ways and manners; however, the richness of the Islamic architecture of such spaces is something incredibly unique.
6. See: http://www.fondationsuisse.fr/FR/architecture2D.html – accessed January 18th 2013.

坦格尔伍德2号住宅

Schwartz/Silver Architects

　　该住宅的设计与其周围伯克夏山区的其他住宅截然不同,但是住宅与周边的环境搭配协调,也满足了主人的需求。住宅坐落在靠山顶的地方,地理位置优越,气势非凡。

　　建筑表皮使用工业材料,装有商用建筑玻璃和阳极氧化铝波纹板。起居空间在约13.7m长的悬挑结构的端部,可以俯瞰外面的美景。该住宅是波士顿交响乐团一位小提琴家的季节性住处。

　　住宅的设计灵感来源于建筑师沃伦·施瓦茨几年前的一趟美国大峡谷之行和当时站在岩崖边看太阳冉冉升起照耀整个峡谷的情景。

　　施瓦茨回忆说:"两年后,我想起我做的一个梦。一天晚上去波士顿交响乐演奏大厅听音乐会时,我把梦里的景象画了出来。梦里出现几个巨大的发出明亮光芒的彩色球体,旋转着、闪闪发光,静静地旋转着跨过天空。有许多人站在那里仰头观望着这一壮观的景象。不久我就开始画住宅的草图,构图是住宅从山边伸出来,这座山就是现在住宅的所在地。"

　　建筑师和他的夫人计划把这里年久破败的老房子拆掉。原来的房子是垂直的,而新住宅是水平的。老房子到处都是墙,景致感觉都是框起来的;新房子大部分墙体安装玻璃,可以看到周围的全景。老房子是木质结构;新住宅是采用钢筋、混凝土和玻璃建造。占住宅50%的悬挑结构对建筑师来讲是结构方面的一项挑战,但是建成的话,住宅会让人觉得有飞翔的感觉。角度会使住宅产生动感。

　　陡峭的选址对于实现住宅的建筑理念至关重要。基本的规划是建一个5.5m×27.4m的长方形,里面有三间11.5m²大的卧室和一间起居空间,起居空间分为座位区、餐厅和厨房,总面积大约45m²。从前门到悬臂的基座拐弯的地方之间,高度大约降了3.8m。伴随这种高度变化的还有三段相同阶数的台阶,形状看起来像鱼梯,楼梯平台通往客房、卫生间和钢板制成的梯子,钢梯从地下室直达屋顶平台。住宅内部石膏壁板上添加了抹灰涂层,室内还装有混凝土地板,走廊边是半

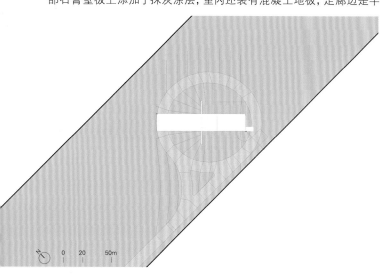

透明的玻璃嵌板和玻璃门。

整个房屋采用的是钢构支架，混凝土厚11cm、钢板屋顶和楼板结构，双倍加固的对角管道支撑。室外包裹了一层阳极氧化铝波纹板，窗框是标准的商用铝质玻璃框。住宅下面的混凝土拱腹和铺设的屋顶平台倾斜3°。

住宅下面现场浇注的地基露出一面2.4m高带洞口的墙，这里可以通往垂直的中央楼梯和机械储藏室。从楼梯中部的平台可以进入屋内。在屋顶，楼梯穿过一面2.4m×3.7m的推拉式屋顶门，直接通往屋顶平台。

Tanglewood House 2

The design of this house has little in common with typical houses of the surrounding Berkshire Hills, but is in harmony with the site and the needs of the homeowners. It is located near the top of a hill, and takes advantage of the location with a commanding presence.

The building skin uses industrial materials, with commercial glazing and corrugated anodized aluminum siding. The living area, at the end of a 13.7m cantilever, opens onto spectacular views. The house is the seasonal home of a violinist with the Boston Symphony Orchestra.

The concept for the house was inspired by a trip that the architect, Warren Schwartz, took to the Grand Canyon (USA) several years ago, and the experience of watching the sun rise over the canyon while standing on a rock precipice.

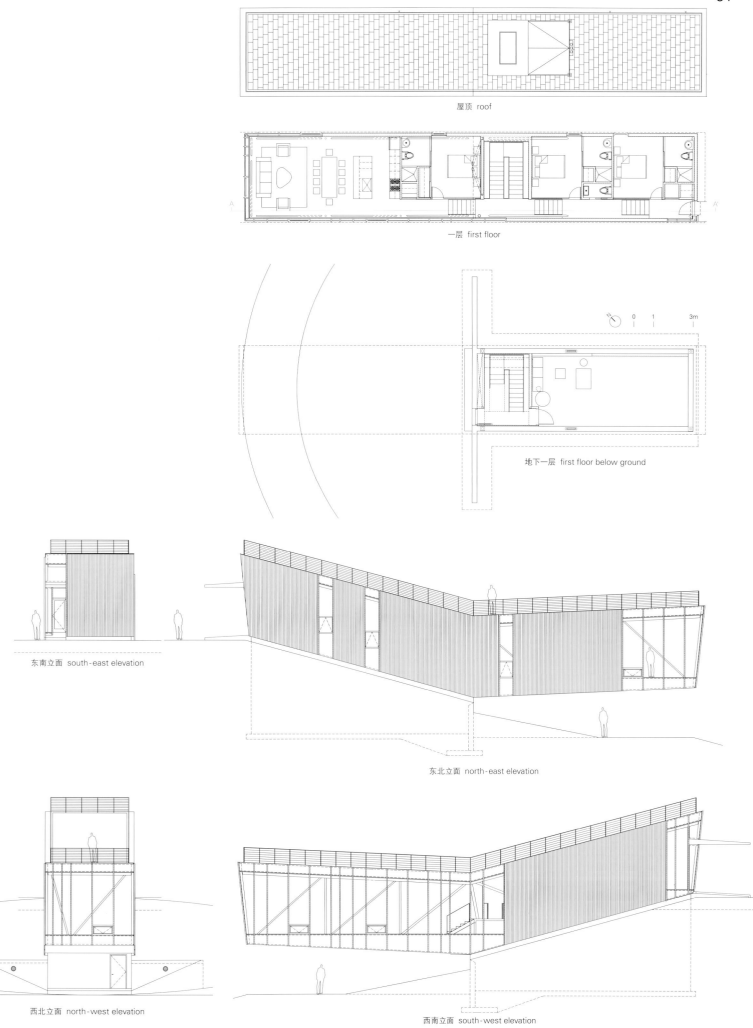

屋顶 roof

一层 first floor

地下一层 first floor below ground

东南立面 south-east elevation

东北立面 north-east elevation

西北立面 north-west elevation

西南立面 south-west elevation

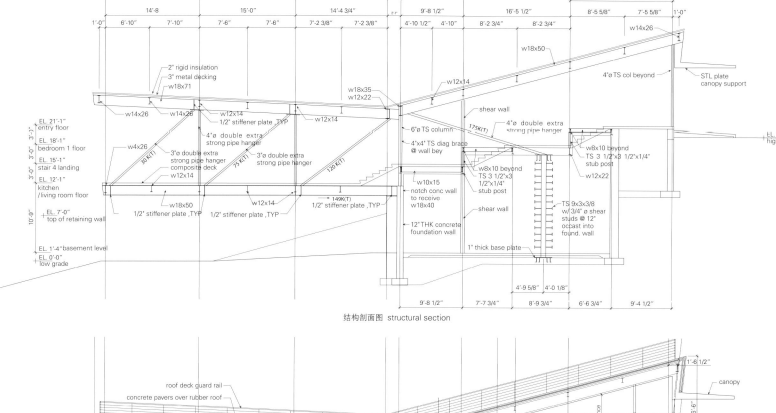

结构剖面图 structural section

A-A' 剖面图 section A-A'

"Two years later, I remember having a dream. I tried to draw it one night while attending a concert at Boston's Symphony Hall. In the dream, there were several enormous spheres brightly lit and colored, rotating and flashing, and silently rolling across the sky. And there were many people standing together and looking up at the spectacular display. Later, I began to sketch a house that projected out from the hillside where my then-present house stood," Schwartz recalls.

The architect and his wife had been considering replacing the previous house because it was weathering poorly. The earlier house was vertical; the new house would be horizontal. The earlier house was mostly wall with framed views; the new house would be mostly glass with panoramic view. The earlier house was made of wood; the new house would be made of steel, concrete, and glass. The 50% cantilever would be a structural challenge, but it makes it seem as if the house could take flight. The angle suggests motion in the form.

The residence's steeply sloped site was fundamental to the conception of the house. The basic plan is a rectangle of 5.5m x 27.4m, containing three 11.5m² bedrooms, and a living space – seating area, dining and kitchen – of approximately 45m². The grade drops approximately 3.8m from the front door to the point of inflection at the base of the cantilever. This elevation change is accommodated by three equal sets of steps, arranging like a fish-ladder, with landings allowing access to guest bedrooms, a toilet room and a steel-plate stair which runs vertically from the basement beneath the house to the roof deck. The interiors are finished with skim-coat plaster over gypsum wallboard, sealed concrete floors, and translucent glass panels and doors along the corridor.

The house is framed as a steel truss, with 11cm thick concrete and steel-deck roof and floor structure, and double-extra-strength pipe diagonal bracing. The exterior is clad in corrugated anodized aluminum and standard glazed commercial aluminum window framing. The concrete soffit underneath the house and the paved roof deck are tilted 3°.

Below the house, the poured-in-place foundation exposes a 2.4m tall concrete wall with an opening allowing access to the vertical central stair and a mechanical and storage room. The stair's intermediate landing opens into the house. And at the top, the stair passes through a 2.4m x 3.7m sliding roof door that allows access to the deck. Schwartz/Silver Architects

项目名称：Tanglewood House 2
地点：West Stockbridge, MA
建筑师：Schwartz/Silver Architects
结构工程师：Sarkis Zerounian & Associates
电气工程师：Sun Engineering, Inc.
施工方：Chris May Builders, Inc.
总建筑面积：148.64m²
竣工时间：2009
摄影师：©Alan Karchmer (except as noted)

Algarrobos住宅

Daniel Moreno Flores + José María Sáez

总长为18m的八个相同的钢构件以X、Y、Z三线为轴,约束着住宅的空间,同时这些钢构件又朝各个方向伸展,使住宅空间具有开放性。这些钢构件像是抽象的梁柱,摆放的位置好像在和周围的环境建立关系。

住宅的结构以它的长体量中和了风景和个人需求两方面因素。巨大的悬臂朝向深谷,还有高耸的桅杆让住宅引人注目;在住宅的狭窄方向上有两根使用高度相同的梁。

住宅采用钢质支撑结构,次级木结构实现了对空间的要求,因为建筑的重复和持续性反而减弱了空间的封闭感。梁依次摆放,支撑着楼板和屋顶,给人悬浮的感觉。

玻璃板起到了保护木材的作用,也使整个封闭的结构变得完整。多数情况下,这些玻璃板是可以挪动的,不论是因为透明性抑或是反射作用,都起到了加强和外部联系的作用。金属屋顶的设计采用反光池是因为建筑师坚持要让池子倒映出周围的环境,一定程度上弱化住宅的存在感。

住宅使用者和场地之间的关系是项目设计的出发点。建筑师致力于加强使用者和现实之间的联系(包括场地、材料、活动),通过简洁的形式和结构体系来实现上述愿景,这种体系也反应了建筑师最初的构想过程。

体系

环境、功能和用户是建筑师设计房屋的出发点,形式和结构构成了具体的方案。当形式和材料的选择在设计过程中相互交织时,建筑师采用了特定的方案,这种方案既连贯又富于变化,既抽象又真实。利用有限的构件,借助连接原则,建筑师建造的不仅仅是一个物体,更是一个体系。通过简化和系统法,建筑师筛选必要的构件,同时加强建筑和外界的关系。建筑师致力打造具有世界性的建筑,可以增强与环境之间关系的建筑。

Algarrobos House

Eight identical steel members, 18 meters long, placed along the XYZ axis, confine the space for the house and at the same time open it as they project in different directions. They are abstract beams that are oriented looking for their relationship with the surroundings.

This structure grants a necessary intermediation between landscape and individual: in its long dimension. It generates the large cantilevers towards the ravine or the masts that mark the presence of the house. In its short dimension two beams are equivalent to one usable height.

Supported by the steel structure, a subsystem of wooden members completes the spatial definition, diluting by repetition and consistency its condition of enclosure. Sequences of equal beams allow the floors and roofs to be sustained or suspended.

Glass planes protect the wood and complete the enclosure system. These planes, in many cases movable, reinforce the relationship with the exterior, be it by transparency or by reflection. The use of reflecting pools on the metal roofs insists on the strategy of mirroring the surroundings, diluting in part the presence of the architecture.

The connection of the user with the place is what generates the project, architects seek to intensify the user's connection with reality (place, material, activity) through a basic formal and construction system that at the same time reveals the origin of their thought process.

System

Environment, function and the user are the starting points that drive the design. Form and construction work out a concrete solution. When formal and material decisions in the design process become interrelated operations, they obtain a specific solution which is at the same time coherent and viable, abstract and real. More than an object they generate a system that is determined by a limited number of elements and rules of conjunction. By decanting elements through simplification and systematization and simultaneously enhancing their ability to generate relationships, architects seek an architecture that is universal, one that intensifies the relationship with environment.

Daniel Moreno Flores + José María Sáez

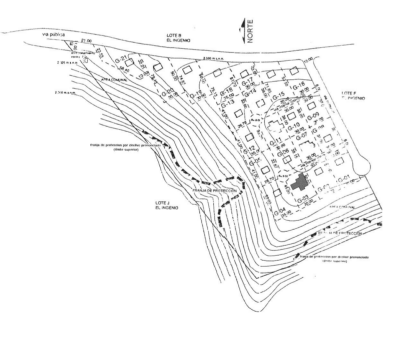

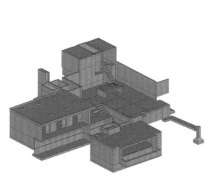
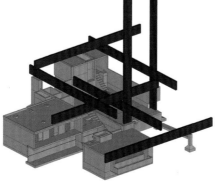
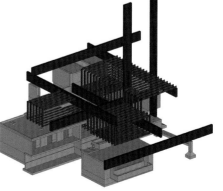

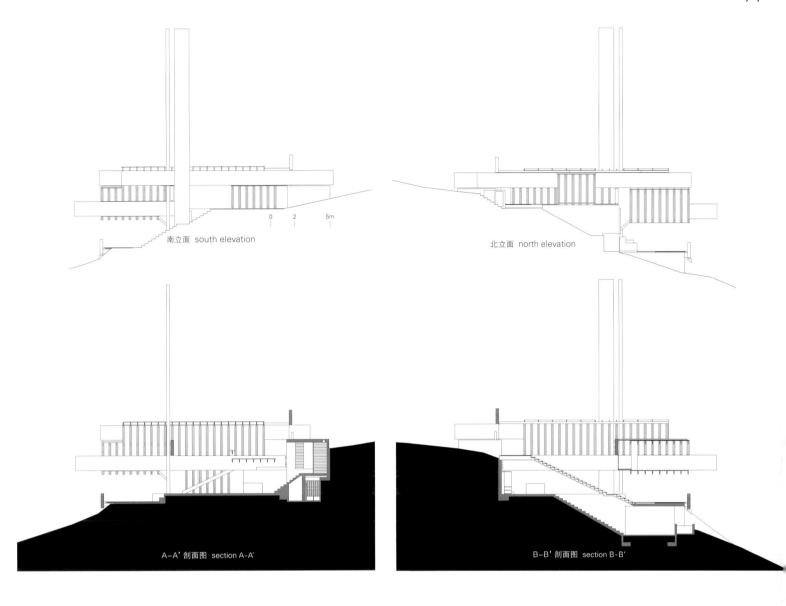

南立面 south elevation

北立面 north elevation

A-A' 剖面图 section A-A'

B-B' 剖面图 section B-B'

项目名称：Algarrobos House
地点：Ecuador, Puembo, Calle del Bagazo, Lote G3
建筑师：José María Sáez, Daniel Moreno Flores
合作者：Margarida Marques, Estefanía Jácome, Santiago Vaca, Claudia Ponce, Estefanía Luna, Adrián Beltrán, Joe Jivaja, Dennise Paredes, Valentina Benalcazar, Amaya Navarrete
工程师：Ing Cesar Izurieta
结构工程师：Cesar Izurieta
电气工程师：Richard Medina
承建商：Luis Guamán
甲方：Paulina Romo and Boris Aguirre
基地面积：1,456.65m² / 总建筑面积：550m²
材料：steel, concrete, wood, glass / 竣工时间：2011
摄影师：courtesy of the architect- p.73, p.74top
©Raed Gindeya Muñoz (courtesy of the architect) - p.71, p.74bottom, p.75
©Sebastian Crespo (courtesy of the architect) - p.66~67, p.68~69, p.70, p.72

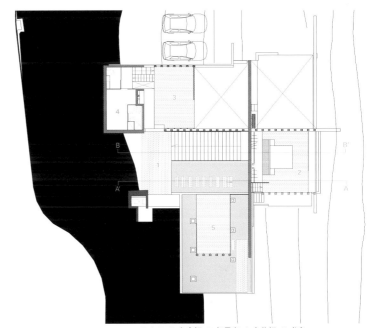

屋顶 roof

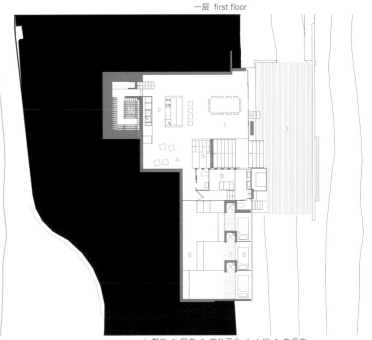

1 入口 2 主房间 3 起居室 4 杂物间 5 书房
1. entrance 2. main room 3. living room 4. utility room 5. study room
一层 first floor

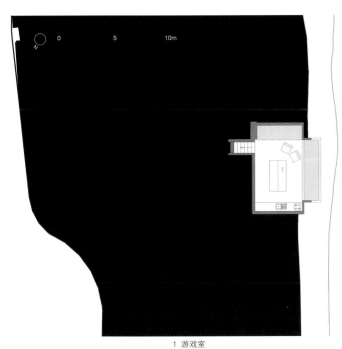

1 游戏室
1. game room
地下二层 second floor below ground

1 餐厅 2 厨房 3 室外平台 4 大厅 5 电视房
6 儿童卧室 7 淋浴室 8 儿童小屋 9 游泳区 10 酒架
1. dining room 2. kitchen 3. outdoor terrace 4. saloon 5. TV room
6. kids' bedroom 7. shower room 8. kids' cabin 9. swimming area 10. wine rack
地下一层 first floor below ground

BF住宅

OAB – Office of Architecture in Barcelona+ADI Arquitectura

飞翔的住宅 Flying House

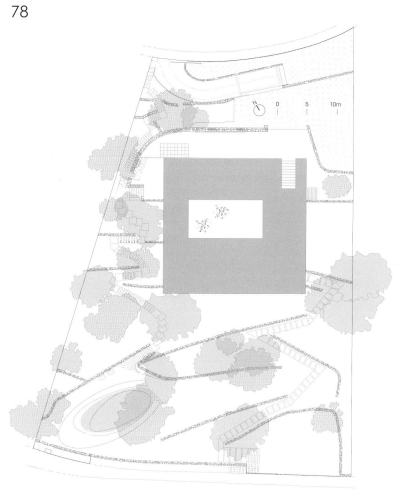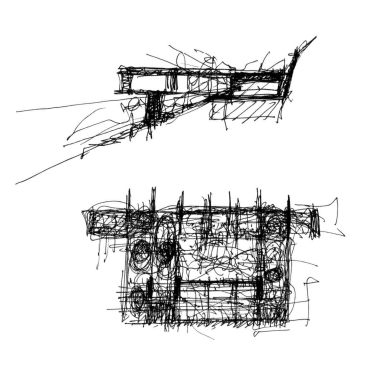

这座住宅高25m、占地3000m², 位于只开发了一半的卡斯特伦社区。稍加观察便不难发现这块土地反映了17世纪的历史, 当时人口过剩的问题迫使人们开发各种地形的土地, 甚至非常陡峭的地形也不例外, 用岩石围墙圈起一系列的阶地。后来这片土地被遗弃, 树木得以生长, 主要是松树和角豆树。建筑师对这片土地的立场是完全尊重的, 所以采用的建筑方法也尊重这片土地, 因此建筑师采用直接放在上面的预制构件的建筑体系, 基本上不动土、不砍树, 利用现有的阶地或花园部分, 只重建受损的部分, 修补中使用相同的石材和技术。住宅的某些部分(车库和辅助区域)被填埋了, 使得建筑师可以在天然的阶地上重新种植当地的草木。这样一来, 叉车可以从坡道靠近建筑的上层。从坡道可以进入位于入口层面下方13m的车库。入口层把建筑的两个层联系起来。这些都在视线之外。说到施工, 为了减少对土地的影响, 建筑师选择工厂预制的金属结构框架, 运到施工现场的大块构件可以安装在3根V形的金属支柱上。现有的石质阶地支撑着整个结构的后部。由于采用了干式建造材料, 整个住宅看起来好似悬在空中, 也像在空中飞翔。住宅的立面分为多个层次, 外面包裹着波纹金属板, 专门用来遮挡眩光和隔热, 这多亏了折叠形成的褶皱。住宅前面开阔的洞口面向壮丽的风景, 在冬季保证充足的阳光进入室内, 夏季也可以遮挡阳光。屋顶安装的太阳能板带热水管道技术, 保证住宅一年四季任何时候都有热水, 满足家庭用热水的同时也可以满足地热所需。气流穿过天井, 利用了不同的朝向, 减少空调的能源消耗, 虽然如此, 住宅还是安装了空调。

住宅中间的庭院可以通往住宅下面, 同时使每个房间都朝阳, 可以欣赏到风景。整个住宅都以这个庭院为中心。这是一座带庭院的住宅, 但在内涵方面却有其特殊性, 因为在庭院的中心可以看到每个房间, 也可以看到周围的风景。虽然庭院的四周是住宅, 但是因为住宅建在斜坡上, 所以庭院未被住宅封闭。住宅前部空间开阔, 有厨房、客厅和主卧, 此处的建筑体系也很清晰, 因为可以看见波纹板支撑的金属支架构成的柱子和屋顶结构, 屋顶架设在波纹板上。

BF House

This home is on a plot of 3,000m² with a height of 25m, in a Castellón neighborhood that is only 50% constructed. Looking at the plot, we see that it reflects its seventeenth century history, which is when overpopulation forced the cultivation of all types of terrain, including those that are very steep, through a system of small terraces with walls made of local rock. The later abandonment allowed the growth of trees, mainly pine and carob. Architects position towards the plot was that of absolute respect, so the construction method should also respect the land, thus opting for a prefabricated building system that is deposited on the land practically without touching it, without cutting down trees, and taking advantage of existing terrace/garden areas, which were rebuilt in the damaged areas, with the same stone and same technique. Part of the house – garage and auxiliary areas – is buried, allow-

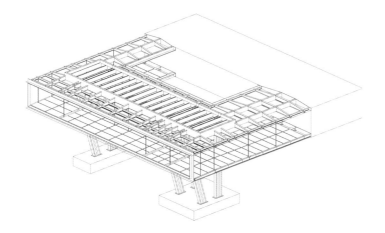

项目名称：BF House
地点：Urbanización La Coma, parcelas 143,146, Borriol Castellón
建筑师：Carlos Ferrater Lambarri, Xavier Martí Galí, Carlos Escura Brau, Carlos Martín Gonzalez
合作者：Benjamin Caballer
用地面积：3,265m²
总建筑面积：619.81m²
施工时间：2006.4—2011.1
摄影师：©Joan Guillamat (courtesy of the architect)(except as noted)

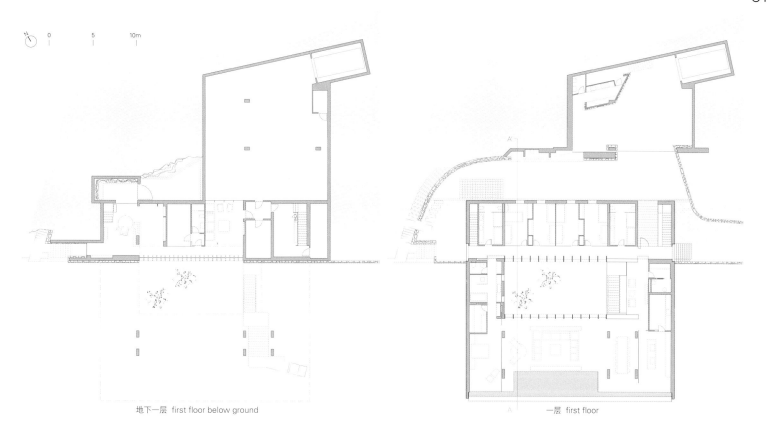

地下一层 first floor below ground   一层 first floor

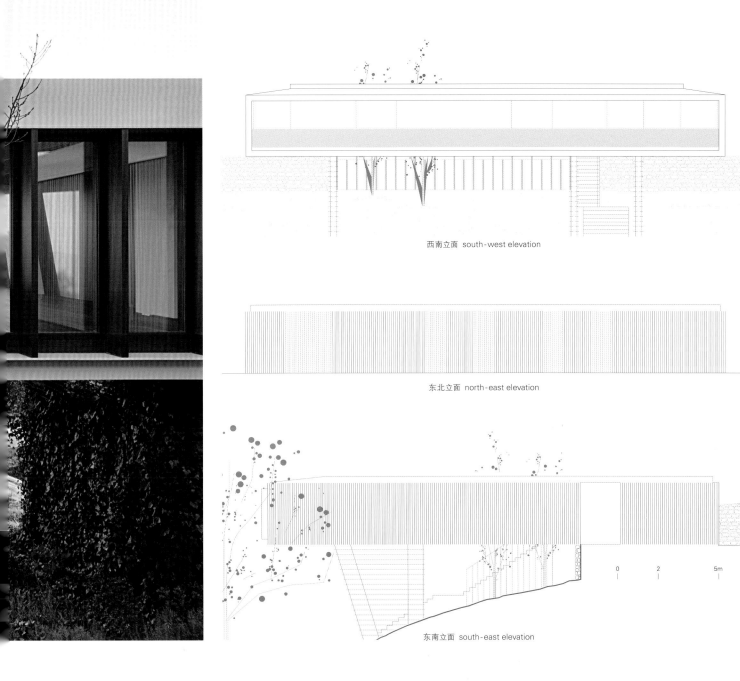

西南立面 south-west elevation

东北立面 north-east elevation

东南立面 south-east elevation

0　2　5m

详图1 detail 1

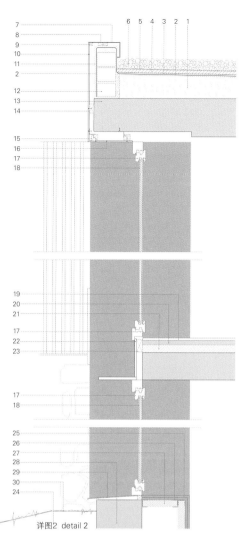

详图2 detail 2

1. cellular concrete
2. waterproof membrane
3. geotextile 150
4. XPS thermal insulation
5. geotextile 300
6. gravel
7. 3mm metal sheet
8. ∅60x20
9. ∅20x20
10. 3mm sheet
11. pur thermal insulation
12. brick wall
13. concrete slab
14. ∅30x60
15. ∅40x40
16. 3mm metal lintel
17. wooden frame
18. glazing
19. wooden pavement
20. self-leveling mortar
21. heated floor system
22. 8mm metal profile
23. ∅40x80
24. soil
25. double plasterboard
26. metal profile
27. thermal-acoustic insulation
28. concrete wall
29. roof railing 3mm metal sheet
30. vegetable layer

ing them to re-introduce native vegetation on the natural terrain. This also allows the plot to be accessed on the upper level by forklift from a ramp that enters the garage located 13m under the access level that communicates with both of the home's levels. All of them are hidden from view. For construction, in trying to lessen the impact on the ground, architects chose a metal structure fabricated in a workshop and transported to the site in large pieces that could be assembled on 3 metal, V-shaped pillars. An existing stone terrace supports the back part of the structure. This home looks as though suspended or in flight due to the dry construction materials used. The facade, resulting in various layers, is finished on the outside with corrugated sheet metal, specially designed to prevent glare and heat, thanks to the shadows caused by the folds. The great front opening is oriented towards magnificent views, and allows adequate sunlight in during the winter, but also protects from the sun in the summer. The solar energy panels with heat pipe technology on the roof allow the home to guarantee that at almost any moment, there will always be hot water available for both domestic use and for under-floor heating. Air currents cross the patio, taking advantage of the different orientations, which permits reductions in air conditioning consumption, which, in any case, has been installed.

The intermediate courtyard allows access under the house, and at the same time, allows all the rooms to face the sun and the views. The whole house revolves around this courtyard. This is a house with a courtyard, but with different connotations since each room in the house can be seen from the courtyard's central location, as well as the surrounding landscape, and since the courtyard is surrounded on four sides by the house, but is not enclosed by it due to the slope of the plot. In the large front area, which houses the kitchen, living room and master bedroom, the construction system is evident since the pillars and roof structure, formed by metal brackets supporting a corrugated sheet over which the roof is built, can be seen. OAB – Office of Architecture in Barcelona + ADI Arquitectura

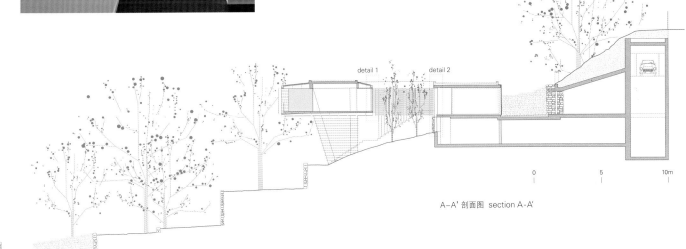

A-A' 剖面图 section A-A'

X住宅

Cadaval & Solà-Morales

造型的力量

　　X住宅项目旨在通过系统的定义、语汇甚至独特的形状来解决一系列的疑问。当看到这个既定的场地时，建筑师在思考如何保护好并且突出一棵令人印象深刻的松树的存在，松树长在场地的最高处，因为这棵松树，从街道进入和接近住宅变得尤其复杂。如何避免两难的选择，住宅到底是侧重海景还是山色，还是在相反的方向分别展示海景和山色？如何通过合适的形式协调现存的不断出现在视野里的建筑物，还是故意遮挡住邻区的建筑？如何成倍突出主要景致，从住宅的前面和后面都能看到美丽的风光？如何通过简单的移动处理好场地上如此多的小修道院的问题，同时实现上述提及的目标但又不厚此薄彼？经过对上述挑战的长久探索，建筑师觉得形式独特的造型是最好的解决办法。这样一来，造型不是依靠先验得出的选择，而是对尝试圆满解决在设计过程中提出的众多疑问的整体回应。

　　X住宅亦是结构上的一种探索：这种技巧其实经常在建桥和挖隧道等基础建设上使用，用在这里是要解决建筑规模的问题，达到提高效率和降低建筑成本的目的。

　　以高密度混凝土为基础的混合技术的使用使得材料可以在高压的情况下放置到单边模板中，并且在极短时间里具有较高的结构阻力。如此一来，6m高的墙就不需要用双边模板了（常规施工方法）。X住宅生动地展示了特定的技术，它的表皮上应用了施工过程中不断积累的知识，这些知识是连续的、多样性的。

　　住宅坐落在巴塞罗那郊区的卡布瑞斯的一座山上。这里风景如画，有一面大斜坡，要到这里只能从最高处仅有的一条街道进来。住宅的选址也考虑到了尽量减少挖掘工作，尽可能优化地使用空地。住宅的通道位于街道下面2m的地方，这个工程试图通过利用空墙实现扎根在这里的愿望，但是从街道上是看不到住宅的；项目优化了住宅的立面和俯瞰峡谷的景致。住宅分两层。顶层除了停车场和进入房子的入口外，也是主人的私人套房，里面有化妆台、洗手间/厕所和宽敞的工作室。在建筑的底层，住宅的前后立面截然不同。住宅前部完全敞开，双层高的起居空间紧挨厨房和餐厅，厨房餐厅里摆放着一张8m

©Santiago Garcés

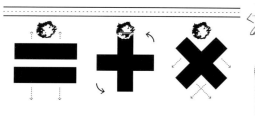
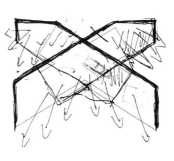
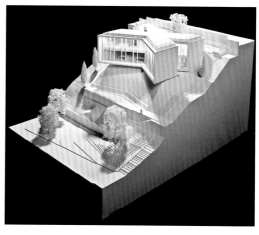

长的大理石餐桌。底层的后面部分设有房间和设备工作区,通过露台可以直接或隐约看到峡谷的景色、大海和山。

X住宅利用造型来诠释空间的不同特质,赋予空间不同的个性特点,并把景观作为主要的元素融入到设计中。

除了住宅前部对空间的合理利用外,每处空间的主角都是景致。运用丹·格雷厄姆的反射手法,欣赏山色的时候不会少了海景,观海时山色倒影在海里:双重的感知特效丰富了住宅的多样体验。

X House

The Power of Form

The X House project aims to solve by the definition of a system, language, or even through a unique form, a number of inquiries that rise up when architects read the specific given site: how to protect and give protagonism to an impressive pine, which is located on the top of the site, and makes access and approximation to the house extremely complex from the street; how to avoid deciding between the views to the sea and those to the mountains, or allow both visions in opposite directions; how to neutralize through forming the presence of the contiguous constructions, to build up a fake isolation that denies the neighbors; how to double the main views, permitting quality frontal views from the front and the rear of the house; how to resolve so many priories with a simple movement that answers to all of the previous aims without prioritizing nor explicitly formulating a response to any of them. The form, a unique form, is the result of a long process of search of individual answers to any of those challenges; thus, the form is not a priori, but an effort to give a unitary response that satisfies each of the questions rose up in the design process.

The X House is also a constructive exploration: a technique regularly used for the infrastructural construction such as bridges and tunnels, is here developed to meet the architectural scale, aiming to incorporate efficiency, and reduction of costs to the construction. The use of a mixed technique based on the application of a high-density concrete allows projecting the material at a high pressure to a single-sided formwork, and to acquire high structural resistance in extremely short periods of time. Thus, it is possible to project continuous 6m high walls without the need to use a two-sided formwork (which would be the regular construction procedure). The house is therefore a living expression of the specific technique, and accumulates in its skin the diverse and continuous knowledge acquired within the process of construction.

The house is located on the upper part of a hill in Cabrils, in the outskirts of Barcelona. The site, with remarkable views and an im-

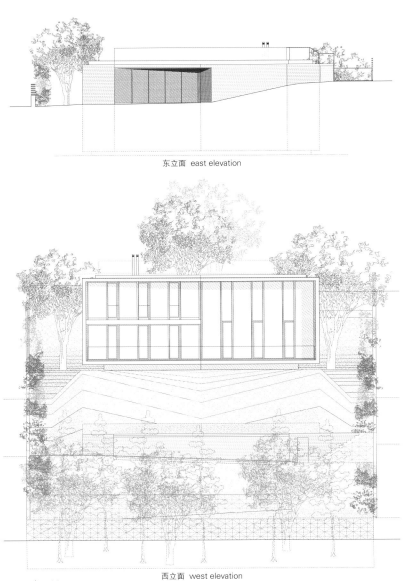

东立面 east elevation

西立面 west elevation

portant slope, is accessed from a single street located at the top of the site. The location of the house within the site responds to the aim to minimize excavation and optimize, within possible, the use of the non-occupied land. The access to the house is two meters depressed from the street, and the project searches to empathise through the use of blank walls the desire to be anchored in the site and to disappear from the street; the project clearly prioritizes the facades and views overseeing the valley. The house has two floors. The top floor, beyond incorporating a parking and allowing the access to the house, is conceived as a private suite of the owners: main room, with dresser and washroom/toilet, and spacious studio. In the lower floor there is a clear distinction between the front and the rear of the house; the front part has a totally open and public nature, building up with a living area in a double-height space next to a kitchen-dining room articulated around a significant marble table, 8m long. The rear part of the lower floor holds the rooms and service areas, which through the patios are given direct and protected views to the valley, the sea and the mountain.

Mainly, the project of the X House uses form to qualify spaces of very different nature and provide them with an individual character, always incorporating landscape as a main actor.

Beyond the effective spatial arrangement at the front of the house, the views are the protagonist in each space. And learning from Dan Graham's reflections, the image of the sea is always present when observing the mountain, and the mountain appears as a reflection when looking at the sea: a perceptive quality that enriches the experience of the house. Cadaval & Solà-Morales

项目名称：X House
地点：Cabrils, Barcelona, Spain
建筑师：Eduardo Cadaval, Clara Solà-Morales
合作者：Bruno Pereira, Pamela Diaz De Leon, Daniela Tramontozzi, Manuel Tojal
建筑工程：Joaquin Pelaez
结构工程师：Carles Gelpi
施工单位：TOPCRET constructions
总建筑面积：300m²
设计时间：2009
竣工时间：2012
摄影师：©Sandra Pereznieto (except as noted)

北立面 north elevation

南立面 south elevation

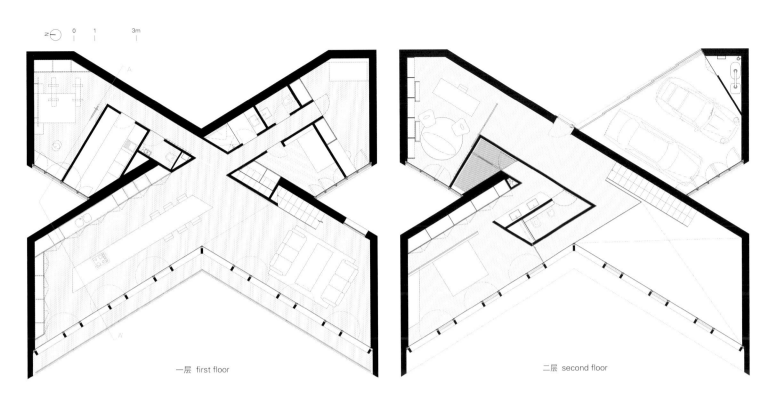

一层 first floor　　　　　　二层 second floor

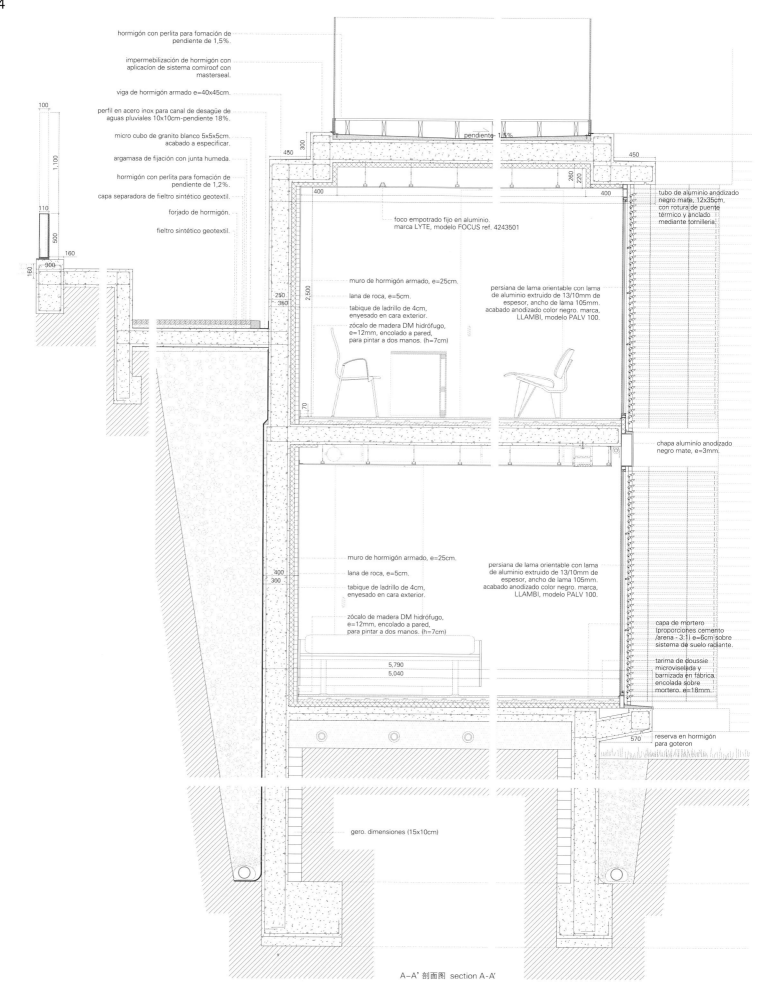

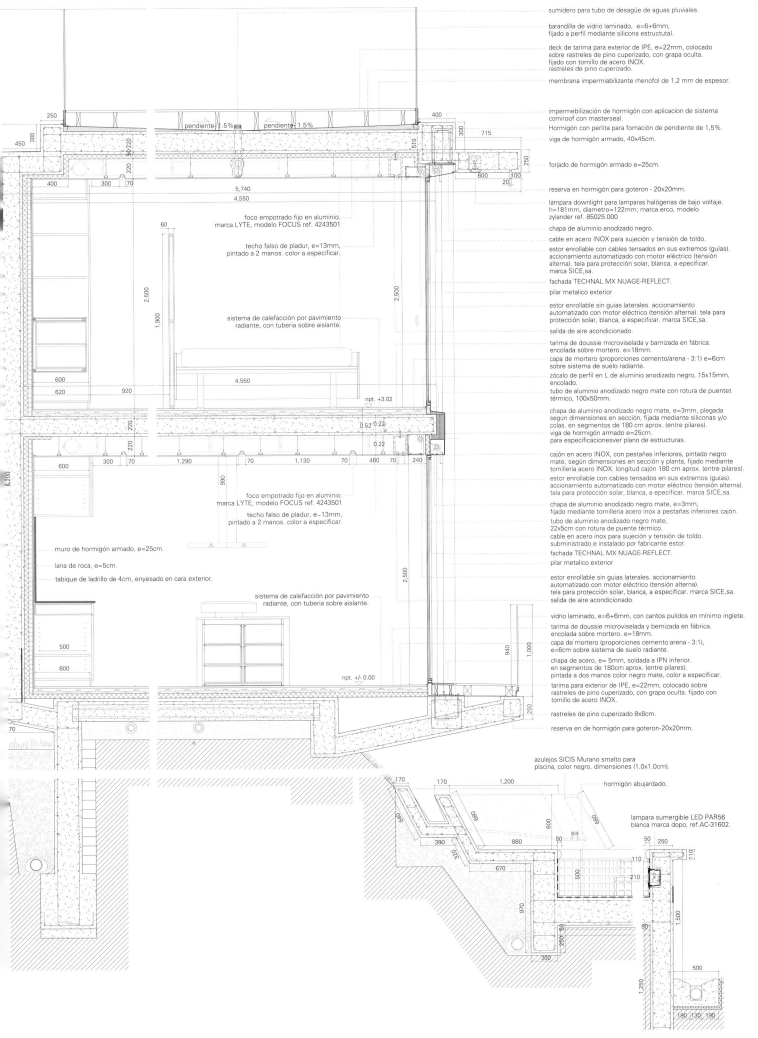

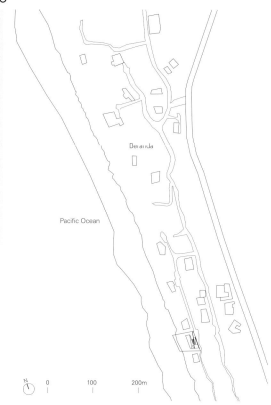

贝兰达住宅
Schmidt Arquitectos

贝兰达住宅位于面朝智利Cachagua海滨和太平洋的一个陡峭的古沙丘上，住宅结构和建筑式样在很大程度上考虑了其特殊的地形条件和底座土壤条件的不稳定性。

住宅建于三面围合的挡土墙上。这些墙把住宅稳稳固定在斜坡之上，形成一系列坡道和平台，成为住宅主体空间不可或缺的组成部分。

一系列坡道和平台沿着主要的流线呈"Z"字形蜿蜒曲折，从公路开始逐渐向沙滩延伸，沟通连接了住宅的各个层面。

木质屋顶依靠一个大型中央钢筋混凝土梁悬吊于主体空间之上，混凝土梁与其下的平面布局相呼应。

住宅由两个关系密切的空间组成。一个是三层高的垂直空间，将建筑各层连为一体，使东面的自然光线可以一直穿透到底层。

另一个是水平展开的宽敞大气的主空间，长35m，包括入口、起居室和餐厅。这个空间是一个大型平台，在此可以领略太平洋的优美风光，满足全家人一切生活所需。

钢框的玻璃幕墙将这一迷人空间的室内和室外区域划分开来，露台和起居室相互交替，滑动窗又可以使两者连为一体。

建筑材料的选用考虑的是如何使住宅与周围环境融为一体。墙面和屋顶交替使用石材和混凝土，保持了建筑与沙丘和海滨上的沙子在视觉上的连续性。

House in Beranda

This house is located on an ancient dune, in an extreme slope front of the beach of Cachagua and the Pacific Ocean. The characteristics of the terrain and the instability of the soil condition to a great extent influence the structure of the project and its architecture.

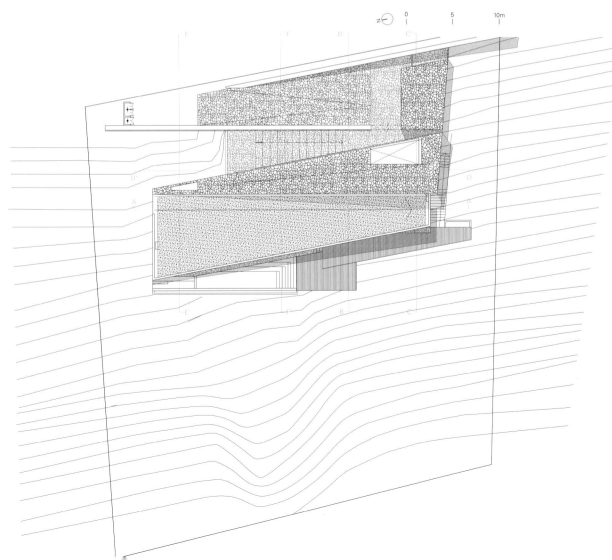

The house is founded on three successive containment walls that anchor the house to the slope and create a series of ramps and platforms that give support to the main spaces of the house.

These are arranged in half floors following a main circulation in zig-zag that gradually descends from the street and communicates all levels.

The wooden roof is suspended above these spaces from a large central reinforced concrete beam that follows the sequence of floor plans under it.

Two relevant spaces articulate the program; one is a vertical space, triple-height, which integrates all levels of the house and allows the natural light of the east to penetrate into the ground floor.

The other is a large longitudinal main space, 35m in length that houses the access, living room, and dining room. This space is intended as a large terrace facing the ocean view and gives place to the family life in all its complexity.

A glass and steel wall divides the outside and inside areas of this great space, alternating terraces and living rooms that can be integrated with each other with sliding windows.

The materiality is designed to integrate the volume with the context. Stone and concrete alternating in walls and roof generate a visual continuity with the sand of the dune and beach.

Schmidt Arquitectos

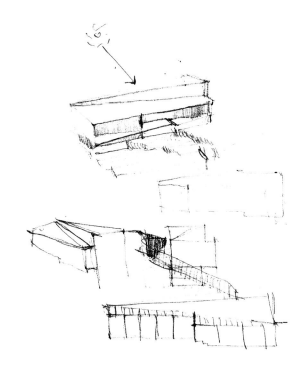

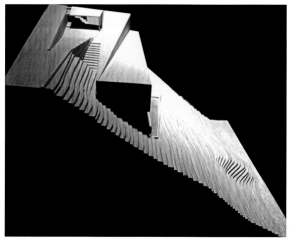

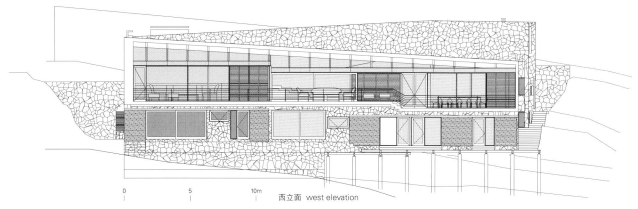

西立面 west elevation

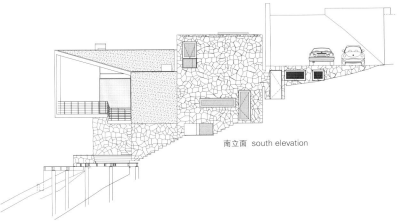
南立面 south elevation
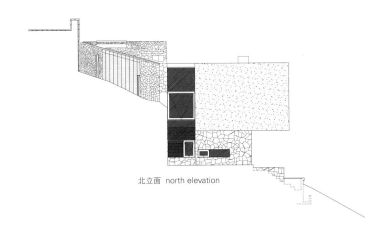
北立面 north elevation

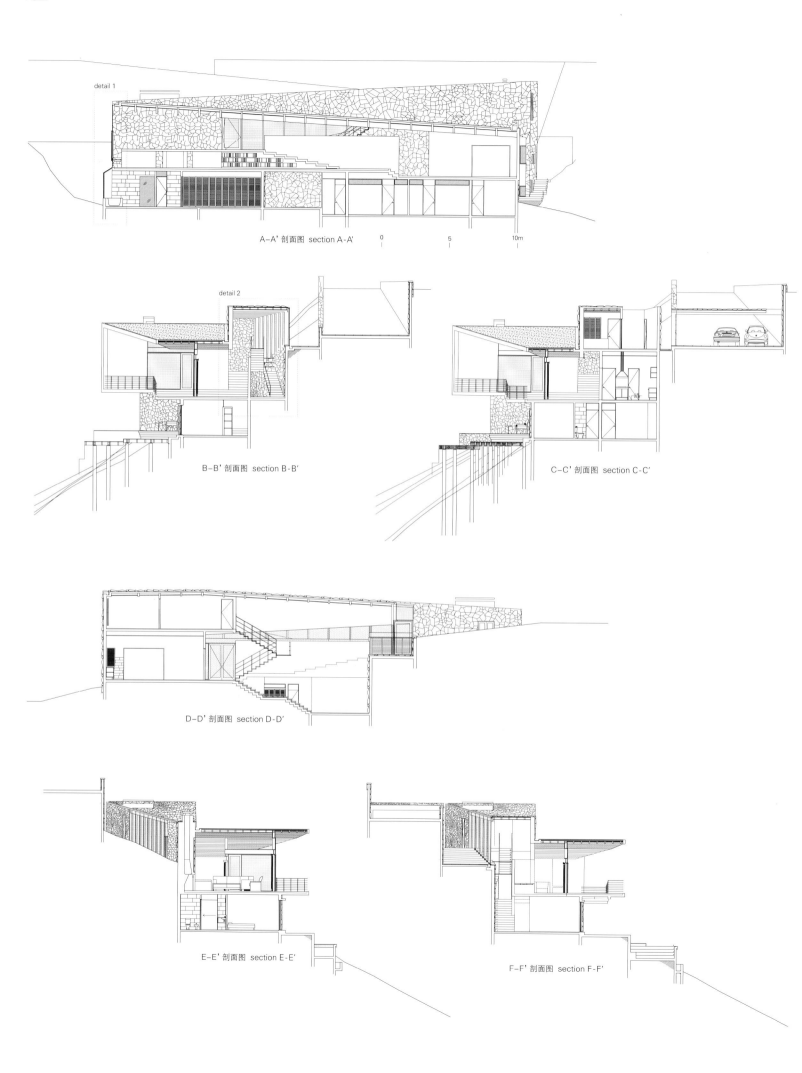

项目名称：House in Beranda
地点：Beranda, Cachagua-Zapallar. V Region, Chile
建筑师：Schmidt Arquitectos
合作建筑师：Paloma Sánchez
结构设计：Valladares Pagliotti and Assoc.
室内设计：Somarriva-Dabove
承包商：Almar.
总建筑面积：420m²
设计时间：2008
施工时间：2009—2011

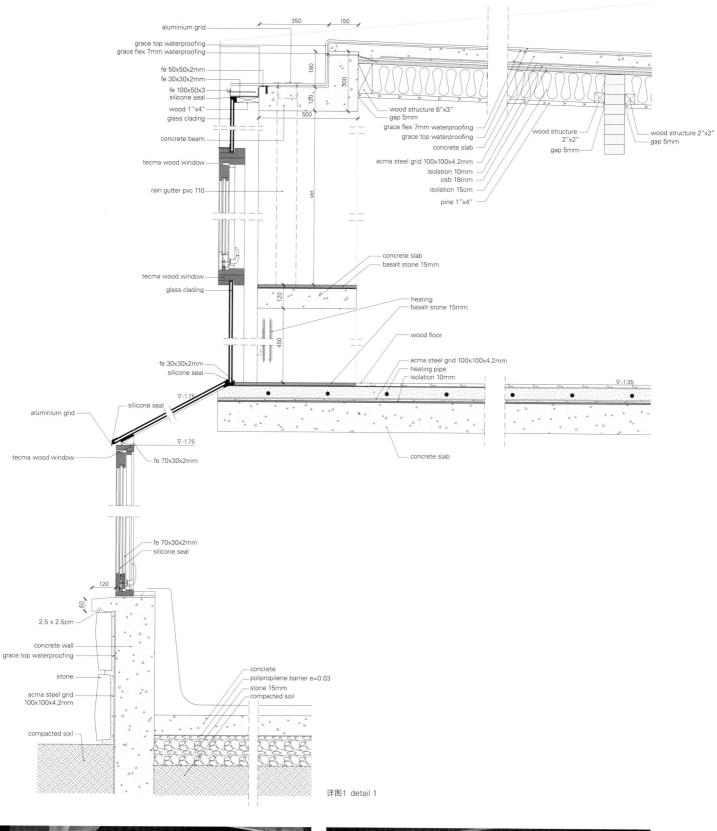

详图1 detail 1

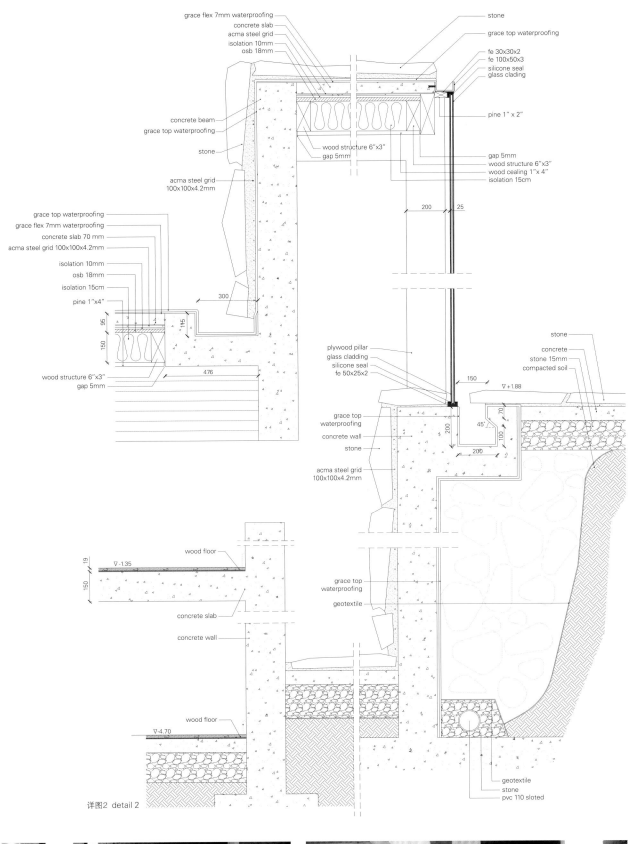

详图2 detail 2

素风宅

acaa/Kazuhiko Kishimoto

这座住宅建筑位于悬崖的半山腰上,鸟瞰大海,山坡上茂密的树丛环绕着住宅,投下郁郁葱葱的树影,似乎让整个山坡活跃起来。本案建筑师认为,此处适宜修建的建筑形式的层高尽可能低,使建筑与地形轮廓相适应,嵌入周边的景观之中,将建筑对环境的影响减小到最低。墙的设计在创造一座建筑的整体存在感方面起着非常重要的作用,因此,本案设计努力避免住宅外墙阻隔或妨碍人体的移动和人的视线。

住宅周身由玻璃和遮挡屏围绕,使居所的二层有种透明感。细长的屋檐在建筑立面上投下深深的阴影,使建筑的物理存在对环境的影响变得柔和。

为了使居住者在每一层都能够欣赏到室外不同的景致,设计师设计了各种各样的空间组成。一层是石材地面和石膏饰面的混凝土墙,花草树木的影子投射在玻璃内侧的日式纸屏风上。棱角分明的表面和饰面与柔和的日式窗户纸相得益彰。

相反,二层设计成开放的生活空间,可以整体面向大海敞开。一系列宽大的屋檐位于屋外与室内之间,一排柱子将其分成较小部分。沿着楼梯形状的露台向下走,可以渐渐亲近室外风景。划分本层两个不同立面的这部分的功能就像独特的日式门廊(缘侧),可以坐在上面。二层使用了钢筋混凝土结构,空腹桥梁结构使设计师能把又大又薄的屋顶架于其上。支柱由实心铁铸造的方形柱子(对角线长75mm)组成,密密麻麻地排列在木质模块(900mm×1800mm)中。通过建立若干低刚性区域,建筑师能够避免使用托架。

Wind-dyed House

A residential building is located halfway up of the cliff, overlooking the ocean. Thick clumps of trees that grow along the slope of the land surrounding the house cast a series of organic silhouettes that make the slope seem to come alive. Architects decided that the appropriate form to build would be as low-lying as possible, while also allowing the architecture to become embedded in the surrounding landscape according to the contours of the terrain. This would allow them to minimize the impact of the building on its environment. The design of the walls plays an important role in creating the overall sense of presence that a building projects. As such, they also tried to prevent the walls of this house from becoming surfaces that would obstruct or impede movement and sight.

Glass and screens along the enclosed perimeter of the house give the second floor of this residence a certain transparency. Slender eaves cast deep shadows on the facade of the building, softening the impact of the building's physical presence in relation to its environment.

The various components of the building were structured in order to allow the inhabitants to enjoy a different view of the outside on each level. The first floor features a stone floor and concrete walls finished with plaster, while the Japanese paper screens fitted inside the glass reflect the shadows of plants and trees. The hard-

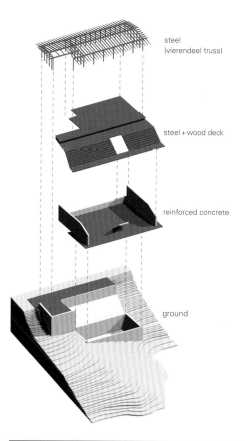

- steel (vierendeel truss)
- steel + wood deck
- reinforced concrete
- ground

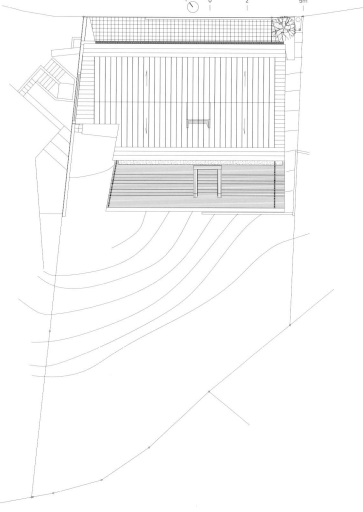

steel + wood deck

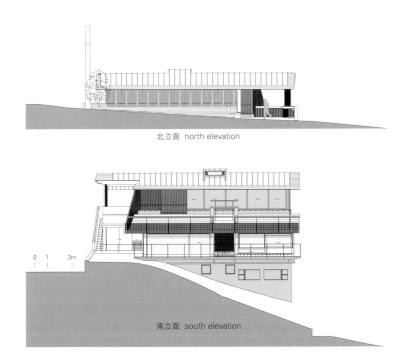

北立面 north elevation

南立面 south elevation

edged surfaces and finishes coexist with the soft, muted tones of the Japanese paper.

The second story, in contrast, features an open-plan living space, the entirety of which can be opened up towards the ocean. A series of wide eaves stand between the outside of the house and the interior, which is articulated into smaller sections by a row of pillars. Going down the staircase-shaped terrace allows one to gradually draw closer to the outdoor landscape. The section that divides the two different elevations on this floor provides seating throughout, functioning as a unique Japanese-style veranda (engawa). A steel-reinforced concrete structure was used for the second floor, and a Vierendeel bridge structure allowed architects to float a large, thin roof on top. The pillars consist of square cylindrical poles (measuring 75mm across) made of solid iron arranged in a densely packed formation using wooden modules (900x1,800mm). By creating several areas of low-level rigidity, architects were able to do away with the need for braces. acaa/ Kazuhiko Kishimoto

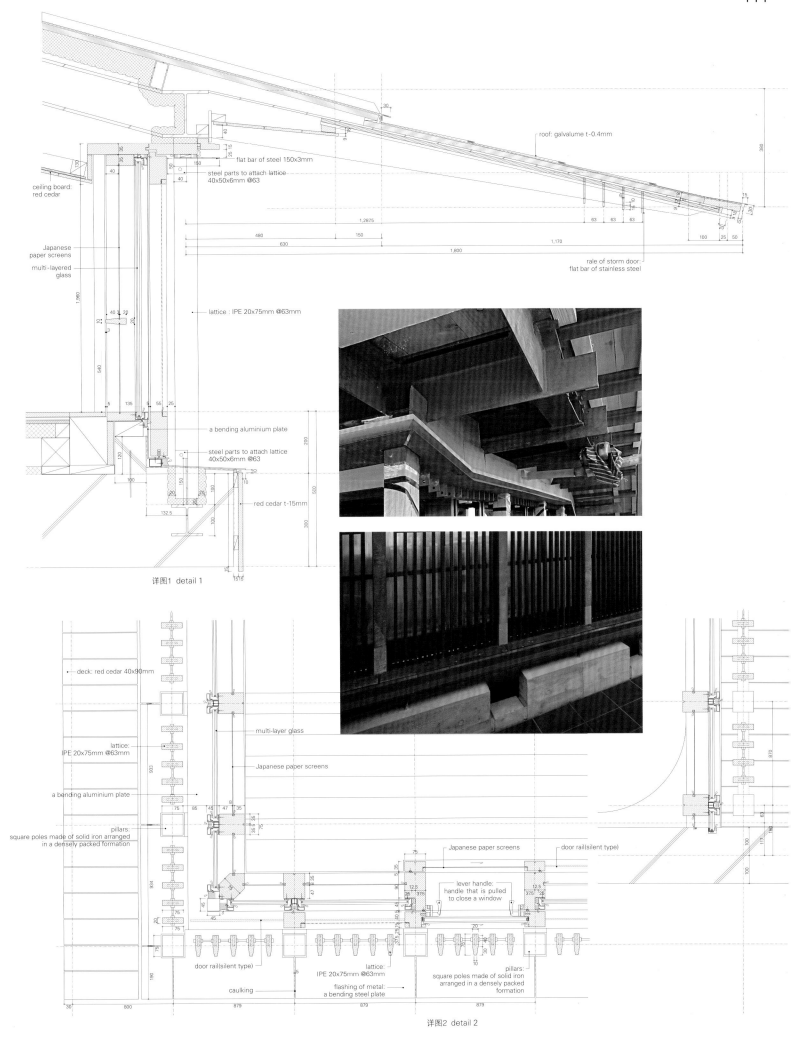

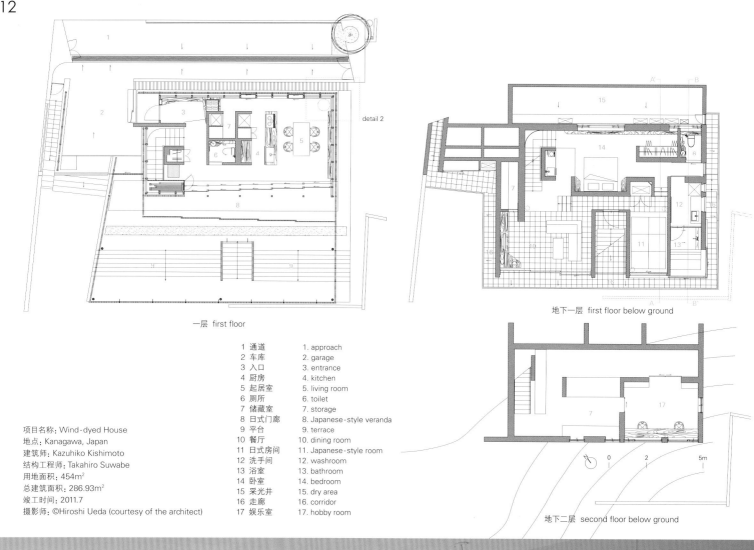

一层 first floor

地下一层 first floor below ground

地下二层 second floor below ground

1	通道	1. approach
2	车库	2. garage
3	入口	3. entrance
4	厨房	4. kitchen
5	起居室	5. living room
6	厕所	6. toilet
7	储藏室	7. storage
8	日式门廊	8. Japanese-style veranda
9	平台	9. terrace
10	餐厅	10. dining room
11	日式房间	11. Japanese-style room
12	洗手间	12. washroom
13	浴室	13. bathroom
14	卧室	14. bedroom
15	采光井	15. dry area
16	走廊	16. corridor
17	娱乐室	17. hobby room

项目名称：Wind-dyed House
地点：Kanagawa, Japan
建筑师：Kazuhiko Kishimoto
结构工程师：Takahiro Suwabe
用地面积：454m²
总建筑面积：286.93m²
竣工时间：2011.7
摄影师：©Hiroshi Ueda (courtesy of the architect)

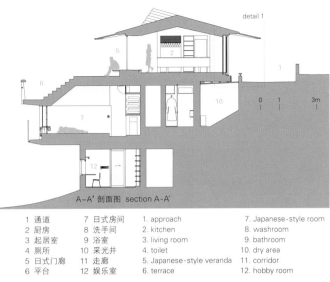

A-A' 剖面图 section A-A'　　　　　　　　　　　　　　　　　　B-B' 剖面图 section B-B'

1 通道	7 日式房间	1. approach	7. Japanese-style room
2 厨房	8 洗手间	2. kitchen	8. washroom
3 起居室	9 浴室	3. living room	9. bathroom
4 厕所	10 采光井	4. toilet	10. dry area
5 日式门廊	11 走廊	5. Japanese-style veranda	11. corridor
6 平台	12 娱乐室	6. terrace	12. hobby room

飞翔的住宅 Flying House

Hanare住宅
Schemata Architects

Hanare住宅是位于日本千叶市的一处隐居之所，项目甲方平时居住在东京，一周只回这座住宅两三次。甲方买下了整座山，本案建筑师负责铺电线、引山泉作饮用水、在灌溉渠上架设一座小桥、房屋的室内外设计以及施工。

项目地点靠近海边，最终选择一座山的西侧斜坡上，场地南面距道路的高差为21m。

房屋及其景观设计得犹如山中城堡，其灵感来自于项目所在地四周群山的轮廓。设计细节与房屋主结构紧密相关。隔墙、设备装置、家具、照明设施和管道等都附着在钢木混合结构的木构件上。简约的设计确保了用户将来希望做任何改动调整时的灵活性。环境影响也是设计中的一个重要因素，长屋檐为内部空间抵挡了强烈的光线和炎热，低辐射玻璃和木质窗框具有极好的保温效果，太阳能吸热墙起到了保持和辐射热量的作用。

Hanare House

Hanare House is a retreat house in Chiba for the client living in Tokyo to use 2 or 3 times in a week. The client bought the whole mountain, and the architects dealt with putting the electricity line, getting drinking water with a fountain, putting a bridge over the irrigation canal, and interior design and architecture, and also construction work.

The site is located near the sea side. It is chosen in a mountain on very steep hill of the small mountain on the west and the south side away from 21m hight from a road.

The house with the landscape is designed more like a castle on a mountain. It was inspired by the profile of the mountains surrounding the site. The details have a close relationship to the main structure. The partitions, fittings, furniture, lighting and pluming are all attached to the timber component of the hybrid steel-timber structure. This simplicity guarantees flexibility for any changes the user may wish to make in the future. Environmental effects are also an important feature. The long eaves provide shade from heat and light; the low-E glass and the wooden sashes give excellent thermal insulation. The trombe walls hold and radiate heat.

Schemata Architecture

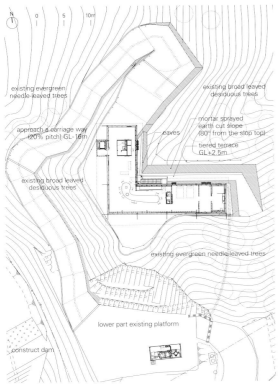

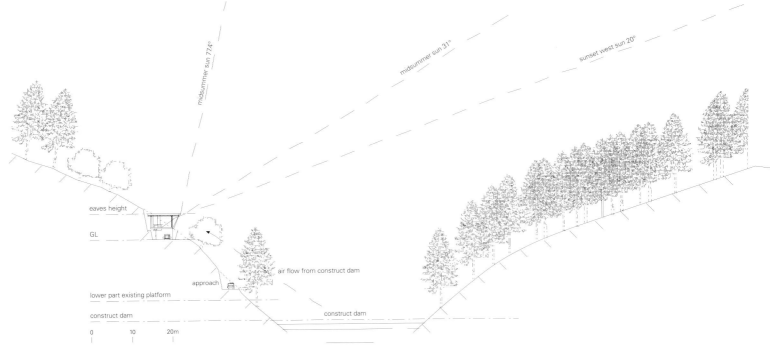

项目名称：Hanare House
地点：Isumi, Chiba, Japan
建筑师：Jo Nagasaka
工程师：E|lrl Structural Engineers
施工工程师：Takaaki Mitsui
用地面积：933.9m²
总建筑面积：180.08m²
结构：steel, wood
竣工时间：2011.10
摄影师：©Takumi Ota

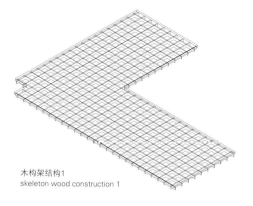

木构架结构1
skeleton wood construction 1

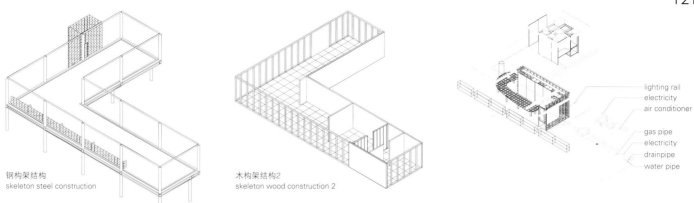

钢构架结构
skeleton steel construction

木构架结构2
skeleton wood construction 2

lighting rail
electricity
air conditioner

gas pipe
electricity
drainpipe
water pipe

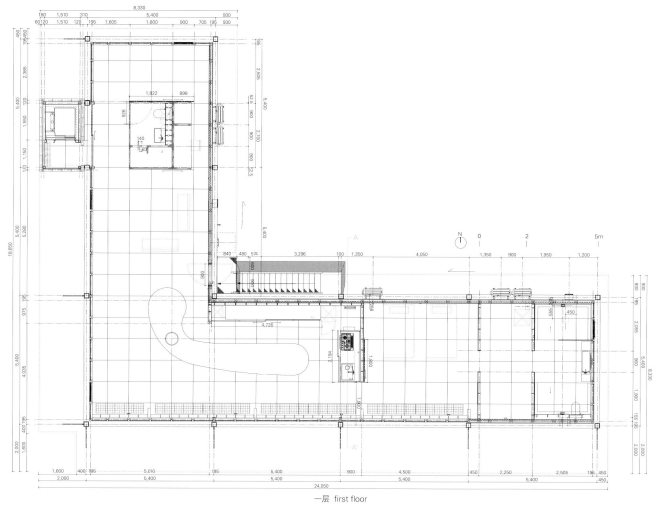

一层 first floor

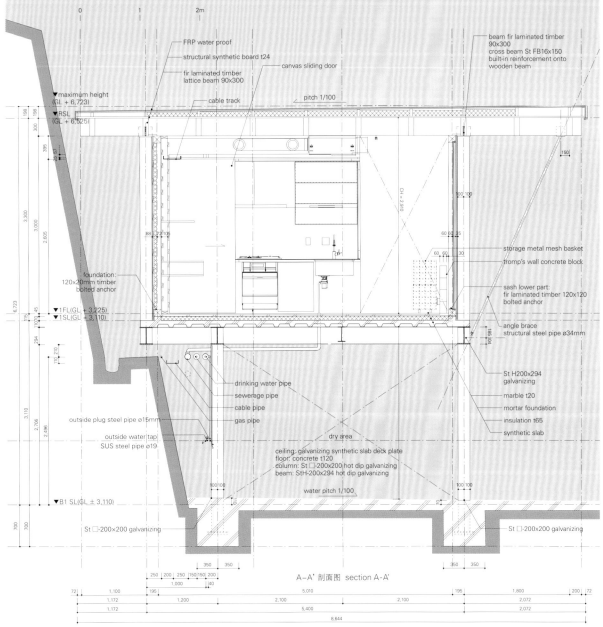

A-A' 剖面图 section A-A'

纳克索斯岛避暑别墅

Ioannis Baltogiannis+Phoebe Giannisi+Zissis Kotionis+Katerina Kritou+Nikolaos Platsas

这座度假别墅是为法国的一个五口之家和他们的宾客所建的，位于一处岩石嶙峋的坡地上，向西望去是碧波万顷的大海。

建筑依斜坡地势分成两座条形建筑，一座主要作为公共空间，另一座包含了卧室和浴室等私密空间。露天区域位于两座条形建筑的中间，包含一个游泳池。泳池是岛上夏日生活的中心。

建筑通过围墙界定了住宅庭院的范围，而场地其他部分则维持原貌，成为周围景观的一部分。

一段3.1m宽的台阶从建筑中间延伸下来，像举办仪式般地引领来访者从停车处进入建筑。台阶身居其中，穿过露天平台的下面，进入私密空间，最终到达一览无余的海景处。巨大的台阶设计气势非凡，使公共平台空间犹如罗马竞技场，向西可以看到海景。此项目修建时，台阶是首先供游客使用的地方。

室内明亮通风，使房屋在夏季的几个月里最大限度地保持凉爽。屋顶种了植物，使建筑本身成为景观的一部分。立面上的开口有明显的节奏交替变化，表现出建筑的理性结构和房屋内部的并列布置。室内的家用设备配置得尽可能少之又少，目的是让居住者更好地体验夏日别墅四周的风景。

Summer House in Naxos

On a rocky inclined site with a sea view to the west, a holiday house has been built for a French family of five and their guests. The building is stepped on a sloping ground in two bars of buildings. One of them is the area in common use and the others are the private quarters such as bedrooms, bathrooms.
The open-air area is laid out on the intervening surface, with the swimming pool, which serves as the center of life in the summer on an island.
The buildings are inscribed within a perimeter wall, which confines residence to its interior and leaves the rest of the site, as a

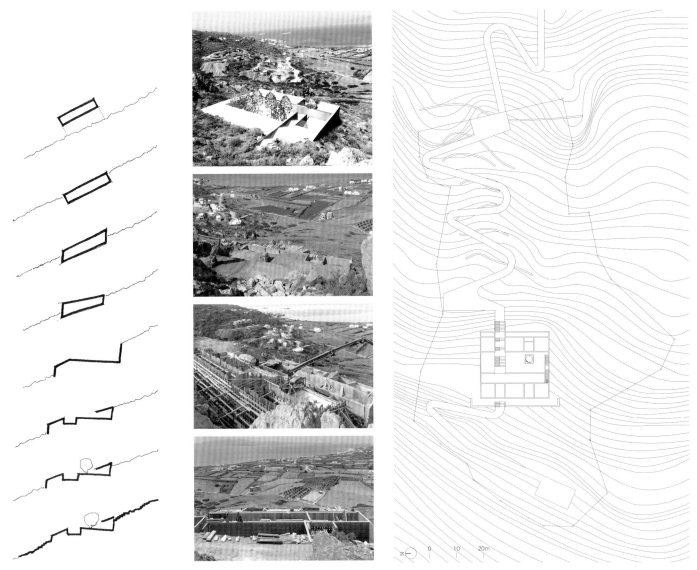

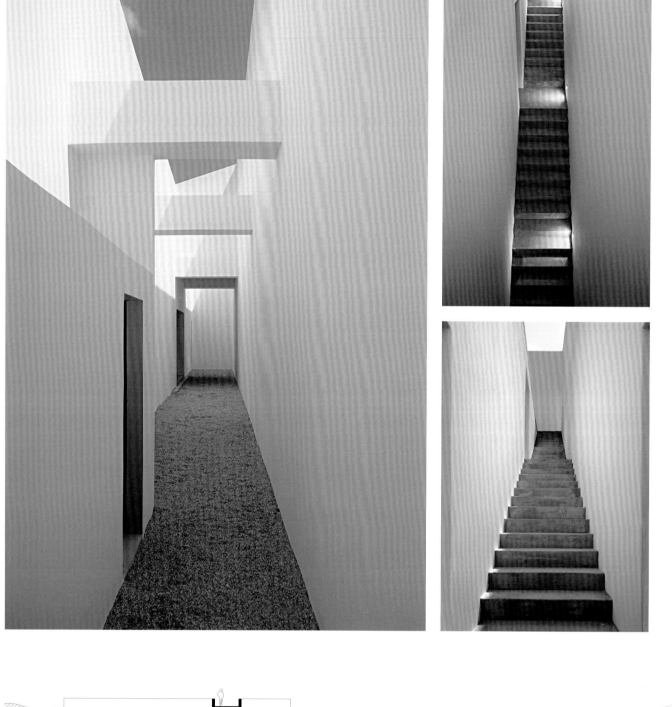
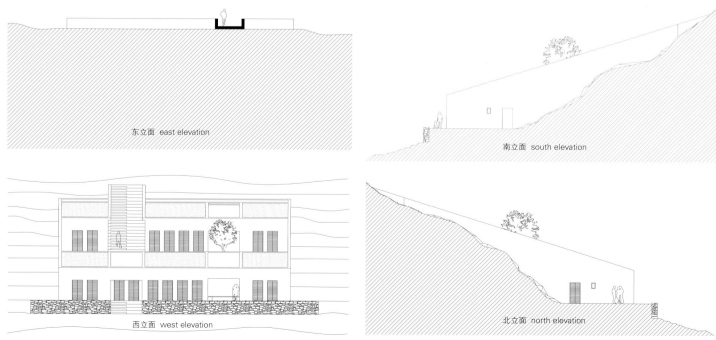

东立面 east elevation

南立面 south elevation

西立面 west elevation

北立面 north elevation

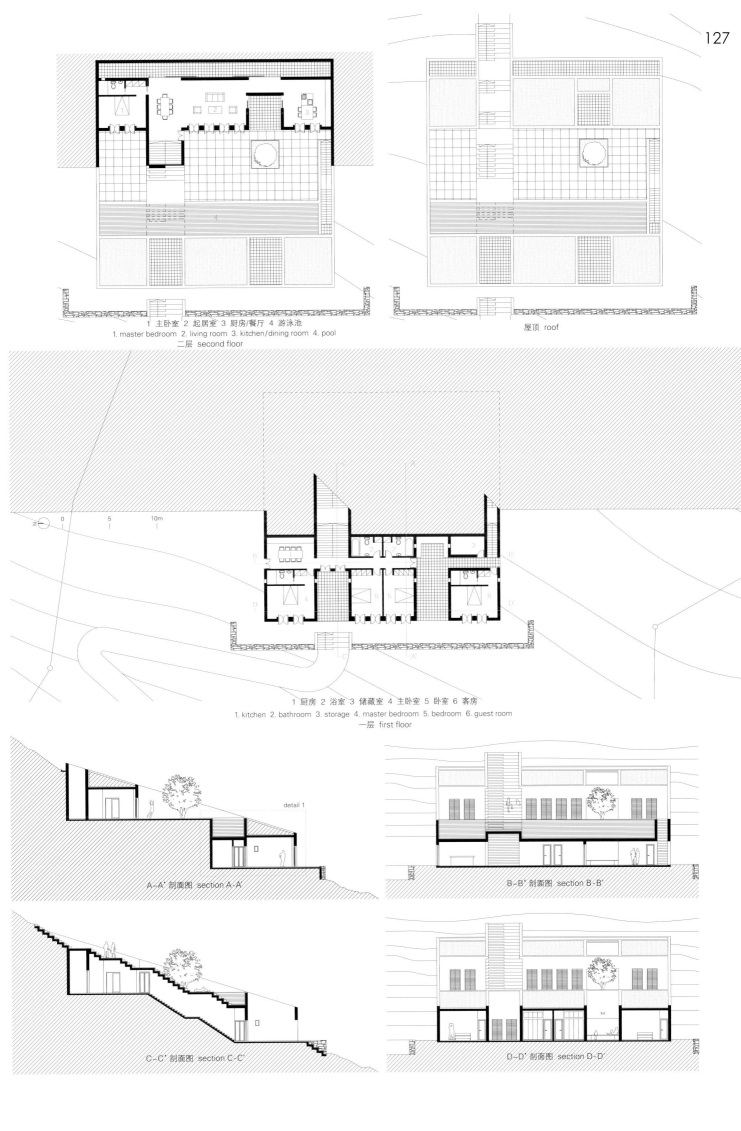

1 主卧室 2 起居室 3 厨房/餐厅 4 游泳池
1. master bedroom 2. living room 3. kitchen/dining room 4. pool
二层 second floor

屋顶 roof

1 厨房 2 浴室 3 储藏室 4 主卧室 5 卧室 6 客房
1. kitchen 2. bathroom 3. storage 4. master bedroom 5. bedroom 6. guest room
一层 first floor

A-A' 剖面图 section A-A

B-B' 剖面图 section B-B'

C-C' 剖面图 section C-C'

D-D' 剖面图 section D-D'

项目名称：Summer House in Naxos
地点：Naxos, Cyclades, Greece
建筑师：Ioannis Baltogiannis, Phoebe Giannisi, Zissis Kotionis,
Katerina Kritou, Nikolaos Platsas
结构工程师：Rodiani Kapiri
总建筑面积：280m²
竣工时间：2012

项目名称：Summer House in Naxos
地点：Naxos, Cyclades, Greece
建筑师：Ioannis Baltogiannis, Phoebe Giannisi, Zissis Kotionis,
Katerina Kritou, Nikolaos Platsas
结构工程师：Rodiani Kapiri

part of the landscape, untouched.

A stairway of breadth of 3.1m passes through the built premises. It brings visitors in a ritual manner from the parking area to the building, distributes them within it, passing below the open-air terrace and penetrating the private areas, and terminates at the open view of the sea. The mass of the stairway creates an amphitheatrical layout on the communal terrace with its sea view to the west. This served as the first accommodation for visitors while the project was being built.

The interiors are all open to the light and air for the maximum cooling of the building during the summer months. The roofs have been planted, and thus the building becomes part of the landscape. The apertures in the elevations with their marked rhythmical alternation give expression to the rational structure and the paratactic arrangement of the premises in the interior of the building. The interiors have been laid out with as little household equipment as possible so that the experience of surrounding space predominates in the activities of residence. Ioannis Baltogiannis +Phoebe Giannisi+Zissis Kotionis+Katerina Kritou+ Nikolaos Platsas

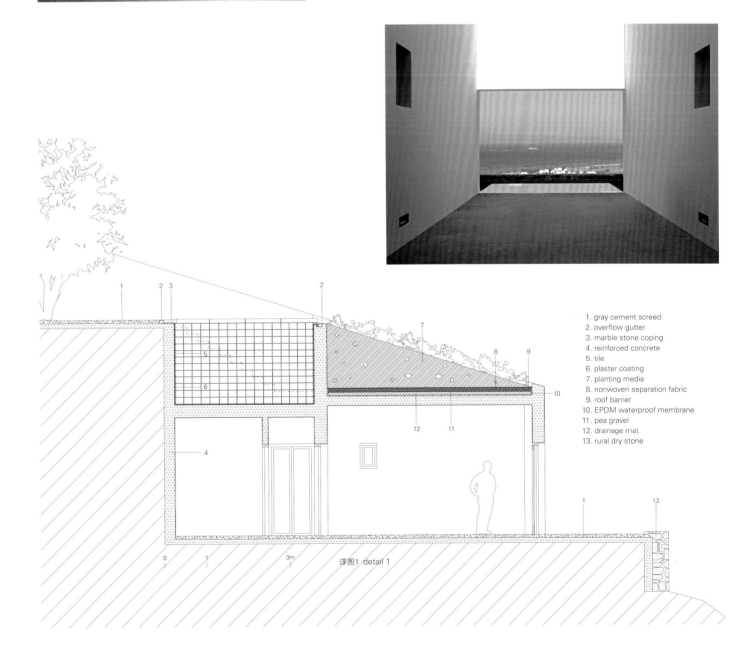

1. gray cement screed
2. overflow gutter
3. marble stone coping
4. reinforced concrete
5. tile
6. plaster coating
7. planting media
8. nonwoven separation fabric
9. roof barrier
10. EPDM waterproof membrane
11. pea gravel
12. drainage mat
13. rural dry stone

详图1 detail 1

Héctor Fernández Elorza建筑师事务所
Héctor Fernández

假若我们相见

下文主要介绍一些个人对建筑的明确观点。有些内容是一些难以用语言表达的建议，而有些是现实存在的、半隐匿在黑暗中却又难以触及的感受。

作为建筑师，总有一些建议与你如影随形。以我为例，我认为有两条建议让我终身受用。第一条是我在马德里高等建筑技术学院学习的最后一年里从建筑师Francisco Javier Sáenz de Oíza那儿学得的。在结束对我们工程项目的修改之后，Francisco Javier Sáenz de Oíza对我们提出了忠告，一座优秀的建筑物就如同一块面包，品尝者不会考虑是谁做出了面包，而只会说："好香的面包啊！"在建筑学中，建筑师（面包师）可能会使其建筑作品（面包）无法实现想表达的主旨思想。Sáenz de Oíza提倡标志不明显的、不受固定模式限制的建筑风格，一如面包。或许，如果建筑风格寻常不起眼，我们可以从另一种方式去欣赏它，建筑作品的内在，理应是无言的表达。

第二条建议来自于建筑师Alejandro de la Sota的一篇文章，名为《优秀建筑让人开怀》。Sota认为，建筑师应成为乐观主义者。他主张，除了注重建筑的外观形象之外，我们还应该聚焦于建筑物所表现的乐观精神，也就是提倡"美好生活"。简而言之，也许是因为如今的社会虚华浮躁之风日盛，"建筑应让人感到幸福"这句老话常常在建筑设计过程中被抛诸脑后了。

在本页两幅手绘图的建筑作品中，存在着两个方面的指导意义：一方面，建筑自身沉默不语，毫无矫揉造作之态；另一方面，从一开始（经常是从画图纸的那一刻开始），建筑就应含有乐观和令人愉悦的特质，让人不禁扬起嘴角。

第一张手绘图是1998年6月绘制的一个虚构的项目，那时我获得研究生奖学金之后前往瑞典入学，刚刚到达瑞典没几天。项目位于

In Case We Meet

The following text is a declaration of intent. Something that is suggested and vaguely hinted at. Something that one knows is there, half-hidden in the darkness but not within easy reach.

As an architect there are statements that remain with you like a good friend wherever you go. In my case I believe these to be two. The first of these came from Francisco Javier Sáenz de Oíza during my final year at the Madrid School of Architecture, ETSAM. On finishing the correction of our projects, he made the comment that good architecture was like a loaf of bread, in the sense that the taster isn't thinking of the baker, but simply says: "what delicious bread!" In architecture, the architect (the baker) can get in the way of the protagonism of his work, from his buildings (his loaf of bread). Sáenz de Oíza proposes an unsigned unlabeled architecture, like a loaf of bread. Perhaps if architecture is anonymous, we would enjoy it in a different way. Behind a work of architecture there should be nothing but silence.

The second statement I took is from a text by the architect Alejandro de la Sota. *"Good architecture makes us laugh"*. Sota suggests that we should be optimists. He argues that beyond the building envelope that shapes the architecture we should focus on fomenting optimism, on promoting "the good life", as it were. In a nutshell, the cliché "architecture should make people happy", is frequently lost, and probably largely on account of vanity.

These are two guiding aspects of the work involved in these two drawings: on the one hand, an architecture is silent and without histrionics and on the other hand, optimistic and cheerful nature with the capacity starts from the outset (often from the very first drawing) to provoke a smile.

The first sketch is from June 1998 of an imaginary project at the

---

贝内西亚公园
双子广场
Valdefierro公园
细胞和遗传生物学学院

材质的高密度 / Jesús Donaire + Héctor Fernández Elorza

Venecia Park
Twin Squares
Valdefierro Park
Faculty of Cellular and Genetic Biology

Intense Material Density / Jesús Donaire + Héctor Fernández Elorza

Elorza

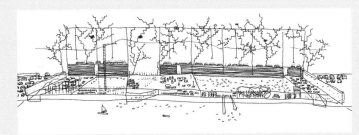

斯德哥尔摩海边一组游泳池的手绘图,1998年6月
sketch for a group of swimming-pools next to the sea in Stockholm, June 1998

西班牙托莱多的山区村舍手绘图,2010年9月
sketch for a cottage in the hills of Toledo, September 2010

Strandvägen街(瑞典斯德哥尔摩市中心的一条林荫大道,意为"海滩大道",译者注)的路边。需要设计的是斯德哥尔摩一个用作停车场的临海广场,我在这里构思了若干个游泳池和一家餐厅。这张手绘图表达了我对20世纪中叶瑞典建筑大师的作品及从中折射出的乐观与活力的由衷羡慕和钦佩。建筑是人类活动的基础,为人们提供支持与帮助,与人们的情感产生共鸣。

第二张手绘图绘自2010年9月,这是一个项目的写生,项目位于西班牙格雷多斯山脚下,是托莱多山中的一个村舍,距马德里一小时车程。房子位于乡村中心,从马德里很快就能到达此地享受假日休闲时光。建造村舍对技术要求不高:建于一片坚固的花岗岩之上,烟囱和电缆塔连接着水井(供暖和淋浴),门上安装了雨篷,桌子用石头砌成,建造塔楼来安放双层床(以达到高度要求),明亮通风的玻璃屋紧挨树梢,从花园里的露台可俯视附近的高原。

这些年里,为了深入思考,我有时候会停下忙碌的脚步,将自己长时间地浸淫于书本和旅行之中,大多数时候都是一个人,不带相机,从内心深处进行反思。

在绘制这两张手绘图之间,我抱着沉静和乐观的生活与工作态度,间或在各处上课、做演讲。

这两张手绘图的绘制间隔十年,其间,我又有了许多梦想,也完成了一些项目。

不断地思索、表达和付诸行动。研究、授课和建设,恰如桌子的三条腿,为我解除疑虑,帮我将想法付诸实施,提供了赖以支撑的稳定性。

注释:在生活中,就像做建筑一样,你可以将时间消耗在咆哮中,也可以沉浸在安静的思索中……

edge of the Strandvägen, a few days after my arrival in Sweden on a postgraduate scholarship. A space in Stockholm occupied car-park next to the sea that I dreamed as a group of swimming-pools and a restaurant. This drawing is the result of my envy and admiration of the architectural works of the Swedish masters of the mid-twentieth century and the optimism and vitality that radiated from them. Architecture is a background to the activities we humans engage in. Construction provides support and resonance to the feelings of Mankind.

The second is a drawing from September 2010: a sketchbook note of a project for a cottage in the hills of Toledo, in the foothills of Gredos, just an hour away from Madrid, where in the heart of the countryside one can enjoy the leisure time afforded by the capital city. Low-tech sensitive architecture: a chimney and a pylon connected to the well (heat and shower), a rocky granite area on which the building is to be erected, doors that are awnings and tables that are stones, a tower of bunk-beds (an excuse for achieving height), a bright, airy, glass-enclosed room next to the tree-tops and the garden terrace overlooking the plateau.

In these years: moments of pause for research. Long periods between books and traveling. Almost alone and without a camera, resulting in much deeper reflection.

Between these two drawings, a few classes and lectures here and there with a limited vocabulary: silence and optimism.

Between these two drawings, in the ten years that separate them, many dreams and some projects.

THINK, SAY and DO. Research, teaching and building like the three legs of the table I lean on providing stability for my doubts and thoughts.

Note: In life, as in architecture, one can spend one's time shouting or in silence... Héctor Fernández Elorza

Héctor Fernández Elorza(左)与Jesús Donaire(右)在HFE办公室
Héctor Fernández Elorza^{left} and Jesús Donaire^{right} at HFE office

材质的高密度

"在过去的几十年里,西班牙已成为建筑品质的堡垒和特色建筑文化综合实力的典范之一。西班牙的建筑依然令人充满信心,集活力四射的传统、牢固的文化根源、明显的社会效应于一体,另外还综合了原因与含义、合理性与情感表达、建筑逻辑与灵感等多重认知与体验。由现代数字技术构思而成的建筑通常会产生陌生的冷漠与距离感,而西班牙建筑则凸显了人类的手工艺技巧。总的说来,在过去的几十年间,缺乏差异性和缺乏主题在建筑设计中日益明显,西班牙的建筑为整个建筑界带来了极大的灵感与希望。"——尤哈尼·帕拉斯马(Juhani Pallasmaa,芬兰建筑师,译者注)

Jesús Donaire(以下简称"JD"):2008年,Héctor Fernández Elorza入选西班牙新生代建筑师代表,参加了西班牙青年建筑师展。在展览目录上,芬兰建筑师尤哈尼·帕拉斯马详细描述了其作品的结构框架,该结构框架产生了普遍却又特别的建筑文化。从这层意义上说,Fernández Elorza以及参与展览的同行们都是建造建筑、研究建筑的建筑师,并始终将教学当成其职业生涯的基本组成部分。这些职能同样是西班牙建筑文化的重要属性,在这种文化的熏陶下,催生了一批注重建筑试验、知识和实践的建筑师群体。再者,Fernández Elorza的作品品质上乘,结构紧凑,能够有效组织利用空间,自身就是功能强大的结构。此外,在我们对空间的感知与对世界的理解上还强调了触觉的重要性。在此背景下,我想知道,您认为自己的建筑在何种程度上属于尤哈尼·帕拉斯马所指的文化,您是否可以为我们详细地描述一下这种文化的主要方面?

Héctor Fernández Elorza(以下简称"HFE"): 近十年来,我一直以建筑为主业,这一时期,西班牙青年建筑师毫无疑问享受到了特别的待遇。尽管西班牙处于严重的经济危机,而这基本上是由于建筑业的过度开发引起的,所幸西班牙政府解决问题的方式还不算太糟。公开竞争体系意味着年轻的建筑师也有机会接触建筑业的全貌,这在其他许多国家都是不可能的。这种极高的参与热情通常是由于缺乏广泛的认可造成的,将朝气蓬勃的西班牙建筑推到了同行业前沿的位置。如果将尤哈尼·帕拉斯马所暗指的文化当成我们的建筑传统、施工与工艺价值的明确属性加以考虑的话,其成果就是传统与现代、现代建筑文化的思索与表现的混合体。

JD: 您在马德里高等建筑技术学院从事教学工作已有十二年了。

Intense Material Density

"During the past decades, Spain has been one of the fortresses of architectural quality and an example of the integrating power of a specific architectural culture. Spanish architecture has continued to project an assuring experience of a dynamic tradition, a feeling of cultural rootedness and societal purpose, as well as a combined sense of reason and meaning, rationality and expression, architectural logic and inspiration. Spanish buildings reflect the presence of human craft and hand compared with the strange sense of cold distance often generated by today's digitally conceived buildings. Simply, during the past decades of growing uniformity and disorientation, Spain has given the architectural world at large inspiration and hope." – Juhani Pallasmaa

Jesús Donaire In 2008 Héctor Fernández Elorza was one of those selected to represent a new generation of architects in the exhibition entitled Young Architects of Spain. Writing in the exhibition catalog, the Finnish architect Juhani Pallasmaa places the works by illustrating this article in their context and describes the precise framework in which they are situated. It's a framework that has engendered a specific architectural culture that is nevertheless universal. In this sense Fernández Elorza, along with his peers who have contributed to the exhibition, is an architect who builds, investigates and maintains his commitment to teaching as a fundamental part of his professional career. These are the same characteristics essential to an architectural culture in Spain that has generated a significant body of experiment, knowledge and practice. Moreover, the work of Fernández Elorza is built qualitatively with an intense physicality that is capable of organizing spaces, of depicting itself as a powerful structure, and which in addition underpins the importance of haptic sensations in our perception of space and our comprehension of the world. Against this backdrop I would like to know to what extent you consider that your own architecture belongs to the culture that Juhani Pallasmaa refers to and whether you could perhaps pinpoint for us the key aspects of such a culture?

Héctor Fernández Elorza During the past ten years, which is the time I have spent working professionally as an architect, the situation of young Spanish architects has clearly been a privileged one. Despite being currently immersed in a deep economic crisis, largely brought about by the excesses of the construction sector, things have not been handled all that badly by the Spanish authorities. The system of open competition here has meant that young architects have access to a panorama that would be impossible in many other countries. This enthusiasm, generally characterized by its lack of excesses and widespread recognition, has placed young Spanish architecture in a vanguard position. If we also take into account what Juhani Pallasmaa alluded to as the clear belonging of tradition in our architecture, the value of construction and craft, the result is a mixture of tradition and modernity, reason and the expression of the present generation's architectural culture.

JD You have spent the last twelve years teaching at the Madrid School of Architecture. How does your teaching influence your work as a professional architect, and vice versa?

苏埃拉古墓入口的手绘图
sketch of the entrance to Zuera cemetery

作为一名专业建筑师,您的教学如何影响您的建筑事业,您的建筑事业又是如何反过来影响您的教学的呢?

HFE: 说到建筑传统,我必须说,建筑传统在很大程度上来自于建筑学院。我曾在马德里高等建筑技术学院学习,现在是这所学院的建筑学教授,我觉得教学在一定程度上已成了我的义务。在学生时代,我的导师Alberto Campo Baeza和Jesús Aparicio对我本科阶段的学习影响深远,他们教会我去体验学校和教学之间的密切联系。在我看来,把从导师那里学到的知识和从自身经历中掌握的知识传授给一代又一代的年轻人,是专业院校主要的职责之一。这种教学观念与过去存在于师徒之间的关系差距不大。导师和学生在工作上密切合作的理念带动了相互交流与沟通的氛围,而年轻人的激情与热情和教师的严格与经验相得益彰。我很乐意告诉我的学生,我从他们身上学到的与他们跟我学到的一样多。他们觉得难以置信,但这却是真的。而且正因如此,知识的交流才更加有意义。

JD: 在翻阅刊载您作品的各类出版物时,人们很容易被这些体现您的建筑理念的手绘图的重要性所打动。这些手绘图通过一些设计元素,诸如明暗度、植被、家具等,在居住和休闲功能方面赋予建筑定性的附加值。我们可以说,所有出现在您的建筑手绘中的这些元素,严格地说,都不算是真正的建筑元素,但它们却定义了您的设计手法,甚至是有意塑造了建筑空间。您能跟我们谈一谈在设计理念形成的过程中最为基础的绘图部分吗?

HFE: 从小我就喜欢画各种器具,对于一丝不苟的手绘过程非常着迷。随着时间的推移,手绘已成为我最喜欢运用的自我表现、情感定位、项目开发的方式。很难去评价我与手绘之间的关系,但是我必须承认,我非常享受手绘的过程,并努力将这种愉悦传播到我的项目中。绘图的魅力在于白纸上的第一条线,不管这条线是一堵墙的第一笔,还是椅子腿或水壶轮廓的第一划,它就如同建筑创作的一粒种子,就在我们全心全意投入绘图的那一刻,这条线在将来会成为一座建筑物或一座公园,那是一种私人的绘画创作瞬间,想象自己就是其中的一分子。

JD: 在您的很多绘图中都出现了建筑物四周的城市框架,从这一点上,能讲讲您对于城市的看法,以及您设计的建筑物、公园是怎样特别根据建筑所采用的材料与周围的大环境形成对话关系的吗?

HFE: 这一问题可以和上一问题联系起来一并回答。给你展示

HFE When we speak of architectural tradition, I must say that this emanates to a large extent from the Schools of Architecture. I studied and am now a professor of architectural projects at the Madrid School of Architecture, ETSAM, and I feel this partly as an obligation. During my student days, my masters Alberto Campo Baeza and Jesús Aparicio, whose influence on me as an undergraduate was decisive, taught me to experience a relationship with the school and teaching as a kind of affiliation. The duty of transmitting to future generations what you have learned from your masters combined with what you have learnt yourself is, in my opinion, one of the principal values of the Academy. This outlook on teaching is not far removed from the relationship that existed in the past between master and apprentice. The idea of working "hand in hand" with students brings about an atmosphere of exchange in which the enthusiasm and fervor of the young blends easily with the rigor and experience of the teaching staff. I tend to tell my students that I learn as much from them as they do from me. They don't believe me, but it is true; besides, this is how such an exchange of knowledge can work meaningfully.

JD On reviewing various publications about your work, one is struck by the importance of your freehand drawings in transmitting your architecture. They provide an added qualitative value to the habitat and the recreational aspect, through elements such as shade, vegetation, furniture, etc. We could say that all those things that appear in your drawings that are not strictly architectural elements qualify your architecture and even intentionally shape the architectural space. Could you tell us about drawing as a fundamental part within the process of generating an idea?

HFE Even in my childhood I was interested in drawing instruments and was fascinated by the precise manner in which they could be used in sketches. Over time sketching has become the means I am most comfortable with in terms of self-expression, defining my emotions and developing my projects. It is difficult to evaluate my relationship with freehand drawing. Yes, I must admit that I enjoy it very much and I try to transmit that enjoyment to my projects. The fascination of drawings that first line on a blank page is enormous and that same line is equally important whether it represents a wall, a chair-leg or the outline of a jug. That is the seed of my architecture, that moment of total absorption in a drawing, which in the future will be a building or a park, that intimate moment of drawing and imagining oneself as being part of the activity.

JD The urban framework surrounding your projects appears in many of your drawings. In this sense could I ask you about your attitude to the city and the manner in which your buildings and parks dialogue with their context, with particular reference to the materials employed in their construction?

HFE I'd like to link this question to the previous one by giving you an example of a specific project and the sketch that gave it shape: it's a drawing that represents the entrance to Zuera cemetery, an intermediate urban place between the town and the cemetery itself. I first drew some high "cold" concrete walls (cold as death); later I imagined some high trees below that would provide shade

1 2 3 4

一个具体的建筑案例及其概念手绘吧：这张图纸描绘的是（西班牙）苏埃拉古墓的入口，是城镇和古墓之间的城市中间地带。起初，我设计了一些高大"冷漠"的混凝土墙（如同死亡一样冷漠），后来，我在墙下增添了一些大树，可以让等候送葬队伍的家人与亲友在树荫下乘凉。这些树木应该是落叶植物，芳香宜人，冬季的阳光穿透枯枝洒落地面，夏季的鲜花为送行的亲友们哀伤的黑色丧服添些暖色。在入口的小广场上修建一条长椅，亲友们可聚集在此小憩，此外还有一个喷泉——"生命之泉"。公墓有几处大门，每一个大门都很高，因为这是世间为数不多的几种让你感觉灵魂超越肉身的地方。门宽仅1m，只能容一人通过，人们不能结伴而过。实际上，在大门上运用了简单的平衡力原理，人通过后门会自动弹回，不需要人"回头"关门——如果没人从墓地逃出就更好了！所以你瞧，绘图和项目开发过程很有趣，我只希望我的热情和激情能体现在自己的建筑中。

JD：在您入行的最初十年中，您很幸运参与了从小型建筑的修复到大型城建项目的开发等各种不同规模的工程，这样的工作经历似乎为规模概念研究以及这种概念的处理方式打下了基础。我相信在您的作品中，无论是在构造方面还是在结构和材料方面，这一主题都在不断出现，在处理这些概念的时候您的想法是什么？

HFE：我对于研究适应人类尺寸的空间以及如何将这种技巧转化到建筑中很着迷。在所有的建筑项目中，我都非常注重大与小的精确定义，无论是大型和小型建筑，还是介于两者之间规模的建筑都一视同仁。这种兴趣是我作为奖学金获得者在罗马西班牙皇家学院求学期间时逐渐养成的。让我给你举两个混凝土建筑的例子吧：米开朗基罗的斯福尔扎教堂，其砌体结构扩展了各种建筑构件的规模，可大可小；位于马杰奥尔圣母玛丽亚大教堂的贝尔尼尼墓地，一条狭长的阶梯通向圣坛，体现了圣坛的重要性。

JD：您谈到您在瑞士正式开始攻读博士学位，又到罗马继续深造，罗马的导师Asplund和瑞典的导师Lewerentz都尤为注重建材的概念。您能详细介绍一下这些研究是如何被用于建筑理念中的吗？在设计项目时，对触觉品质的体现需要到何种程度？

HFE：在导师Asplund和Lewerentz的作品中，最吸引我的方面除了大环境和结构之外，莫过于与材料处理方式相关的主题了。我对这些大师作品中将亲自种植与培养材料作为决定性因素的做法很有兴趣，而这些也都体现在我大部分的作品中。如果建筑在很大程度上能查出你在周边环境中发现的事物之间的距离，那么近在眼前的距离就

and protection to family and friends awaiting the funeral cortege. These trees should be aromatic and deciduous allowing the sunlight to filter through in winter and flowering in summer in contrast with the black clothes of the mourners. A single, long bench was built on the small entrance square for family members to sit on together as well as a fountain "of life", as it were. There are several gates into the cemetery, all of them high, as they are few places quite like this where you feel that your soul passes above you. These gates open out just a bare meter for people to make their way through one by one and not in a group. Actually a simple counter-weight ensures that the gates close by themselves after people have made their way through, so as not to have to "look behind" – it's better if nobody escapes from a cemetery! So you see that the drawing and the development process especially can be fun and I just hope that this enthusiasm and excitement translates into what I build.

JD During the first decade of your professional career you were fortunate enough to work in projects of varying scales, from small rehabilitation projects to large-scale buildings and urban activities. This opportunity seems to have served as a basis for research on the concept of scale and the way this concept can be manipulated, which I believe is a recurring theme in your work both at constructive as well as structural and material level. What is your thinking on this manipulation?

HFE I am fascinated by the manipulation of things in relation to the size of people and how this mechanism transforms architecture. In all my projects I place special attention on the precise definition of large and small and the play that architecture permits between those two terms. This interest is something that increased during the time I spent as a scholarship student at the Royal Spanish Academy in Rome. Let me give you two concrete examples: Michelangelo's Sforza Chapel, where the manipulation of the masonry stretches the scale of the elements making them larger or smaller, and Bernini's Tomb also in the Church of Santa Maria Maggiore where a narrow, long step access step to the altar manipulates the scale of its significance.

JD You speak to us of Rome where you continued your doctoral research that you had formally begun in Sweden. Both Rome and the Swedish masters Asplund and Lewerentz carry the hallmark of a necessary and intense rapprochement to the concept of material. Could I ask you to briefly decipher how that research has translated into your architectural thinking, and to what extent haptic qualities are essential to your manner of designing projects?

HFE Along with context and structure, their involvement in themes related to the manipulation of material is the aspect that most interests me about the work of Asplund and Lewerentz. An interest in cultivating and fostering material for itself as a determining factor in the work of these masters is also very presented in the majority of my own works. If architecture to a large extent traces the distances to things that you find in the surrounding proximity,

1 圣埃斯特万的乡间石墙
2 圣埃斯特万文化中心正在施工中的外墙
3 哈兰迪雅教堂的墙壁
4 施工中的Valdefierro公园
5 细胞和遗传生物学学院的教师办公室

1. vernacular wall in San Esteban del Valle
2. exterior wall under construction in San Esteban del Valle's Cultural Center
3. wall of the Jarandilla de la Vera's Church
4. Valdefierro Park under construction
5. teacher's office in the Faculty of Cellular and Genetic Biology

是被各种材质与空间占据的区域，在触手可及的距离，就让人无法看到墙壁了。

JD: 在建筑语言中，由建材决定的这种重要性或兴趣仍然是您最近作品中的基础部分吗？比如在哈兰迪雅和圣埃斯特万的这些还在施工的项目。您认为有哪些特点是普遍存在于您的作品中的？

HFE: 通常情况下，在哈兰迪雅教堂和圣埃斯特万文化中心的修复工程中，材料的处理都会受到环境状况的限制。在哈兰迪雅项目中，拆除教堂破旧屋顶的木板瓦用来覆盖新混凝土墙。在圣埃斯特万项目中，村民遵循修建田地周围石墙那样的古老传统方法来修建临街的外围墙。在材料的使用选择上永远没有硬性规定，而是根据建筑具体的环境决定使用某种建材。例如，在修建Valdefierro公园时，移除原址的残垣断壁要花费大量金钱，于是，萌生了利用碎石建造混凝土墙的想法。正是最初的想法决定了建筑最终的质感。

JD: 的确，可以说细胞和遗传生物学学院大楼的建设界定了您职业生涯的特殊时刻。这是一个在外形上脱离传统范畴的大胆项目，既沉重又轻盈、既模糊又透明／半透明、既有新建又有翻修，在结构可靠性上呈静态，同时在建筑物及其材料（混凝土和玻璃）与光线和周边公共区域引人注目的相互作用中又呈动态。您在设计中是如何解决这些矛盾、面对由这种模棱两可的特质所引起的挑战的呢？

HFE: 细胞和遗传生物学学院大楼的扩建和修复工程集环境、结构和建材等各方面考量于一体，最近几年一直很吸引我。该建筑与周围环境和谐共处，特别是延至学院整个西边的花园。建筑西面的立面采用玻璃表皮，建筑物外是单行道，地板如地毯一样延至屋外，使研讨室和走廊在视觉上合为一体。在二楼和三楼朝向西侧的办公室里安装了厚重的落地百叶窗，不但能将花园美景尽收眼底，也能遮挡夕阳西下时分照进屋内的余晖。结构大致决定了建筑最终的外观。扩建部分的混凝土、修复部分的结构加筋和原有的结构都清晰可见。整个建筑都没有安装吊顶，因此这种可见性将结构变成了决定建筑空间的基础。最后，饰面和各种设备的材料也都符合建筑物的最大特色——清晰明了。与环境配套的家具为整个工程画上了完美的句号。建筑的结构和建材的性质在人际关系中构造了中性的背景。

nearby distance is the zone occupied by textures and the space of air that separates the eye from the walls at a hand's distance.

JD Is this significance or interest in an architectural language determined by material also part of the genesis of your more recent works under construction such as those in Jarandilla or San Esteban? What are the characteristics that you would define as common to all your works?

HFE The manipulation of material in the rehabilitation of the Church of Jarandilla de la Vera and the Cultural Center in San Esteban del Valle is, as frequently happens, conditioned by context. In Jarandilla, the wooden shingles of the old roof that had to be demolished are being used to cover the new concrete walls. In San Esteban, the outside enclosing walls to the street are done by the village people following the same age-old tradition employed in building the stone walls surrounding their fields in these parts. Decisions regarding the use of material are never imposed but arise from observing the specific context in question. A clear example is that of the Valdefierro Park where the considerable cost of removing the debris of the original site led to the idea of concrete walls filled with rubble. It was that initial decision that defined its final texture.

JD It is true to say that the Faculty of Cellular and Genetic Biology defines a special moment in your professional career. It is a daring, formally unconventional work, heavy and light, opaque and transparent/translucid, new and refurbished, static in its constructive solidity and at the same time dynamic in the dramatic way in which the building and its materials (concrete and glass) respond to light and define its public areas. How does your architecture resolve the conflicts and challenges posed by such ambiguities?

HFE The extension and rehabilitation of the Faculty of Cellular and Genetic Biology summarizes the aspects of context, structure and materials that have interested me in recent years. It responds to a set of clear conditions of context, especially in terms of the gardens that extend along its entire western side. On this side of the building the facade is glass-covered, traffic is restricted to one lane and its floor extends outwards like a carpet bringing about visual continuity from the seminar-rooms and corridors. The offices of the second and third floors open out in this direction thanks to the deep brise-soleils affording views onto the gardens without the discomfort of the late afternoon sunlight. The structure largely defines the final look of the building. The concrete used in the extension, the structural reinforcement of the refurbishment and the original structure are all visible. This aspect, backed by the non-existence of suspended ceilings in the entire building, turns the structure into the support that defines the architectural space. Finally, the materials used for the finishes and the installations are faithful to their most explicit and intimate character. Together with the context, the furniture does the rest. Structure and nature of the materials weave a neutral background to human relationships.

贝内西亚公园

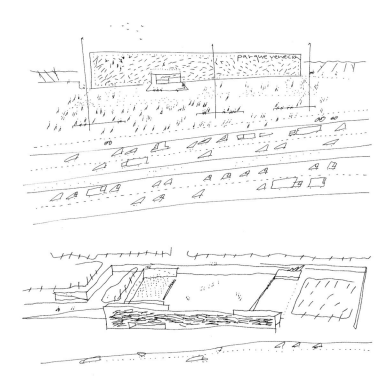

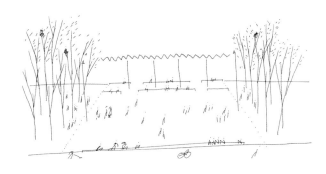

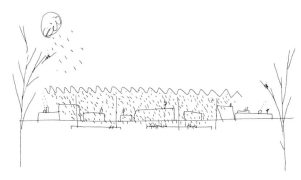

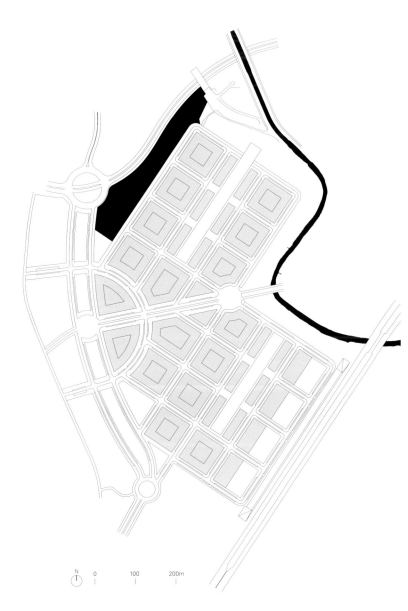

著名的贝内西亚公园是88/1区域内的一片绿地，坐落在该区域的西北端，与Ronda Hispanidad大道平行，位于扎菲罗街道环岛和历史悠久的阿拉贡皇家运河之间。公园平均长415m，宽60m，占地大约2.5公顷，涵盖了一处线性城市基础设施。工程需要解决3个问题：噪音难题、降雨排洪问题和地形问题。

噪音问题是Ronda Hispanidad大道（三环路）上繁忙的交通引起的，影响周围居民的生活，这就要求修建一处隔音带，把公园整个西北边界都包含进去。新建居民区和环路之间现有地形的地面水平高差最大距达到14m。地形的差距依靠加筋挡土墙解决，挡土墙分为四个台阶，每两个台阶之间相隔1.5m，由镀锌钢丝网和大砾石块组成，形成隔音带保护本区未来的居民。

公园西南远端没有明显的地形问题。本案建筑师修建了长100m、最高达10m的高墙解决了噪音问题，这堵墙更是被视为新建街区的一个象征，由此还可通往地下广场或墙体拱背处的磨坊池。

层流池的设计是为了解决影响当地的大量降水，从而防止洪水注入市政管网——市政管网的直径和排水能力不足以处理如此大规模的降水量。层流池的表面积达3150m²，层流空间的规模取决于当地降雨的频率和强度。一年中的大多数时间，这个地方可以是城市空间或步行广场，人们也喜欢到这里来躲避从城市地势高处吹来的恼人的西尔左风。公园四角的四个楼梯通往地下广场，连接毗邻的街区和城市。隔音墙内坡道宽阔，使得服务及养护车辆可以进入，对这一综合设施有了更充分的利用。

贝内西亚公园地形设计缜密，除了为城市提供了绿色空间之外，还包含了上面提到的声学功能及水流形成过程。整座公园建于Ronda Hispanidad大道之上，由线性的或扩展成小广场（硬质或软质）的交错相连的平台搭建而成，观景点处在轻金属棚架、绵延的松树林和人行坡道的保护之中，通过它可以看到阿拉贡皇家运河的历史古迹。

Venecia Park

The green space within sector 88/1, known as Venecia Park, is located at its north-western limits, running parallel to the Ronda Hispanidad Avenue between the Calle Zafiro Roundabout and the historic channel of the Imperial Canal of Aragón. The project encompasses a linear urban infrastructure, averaging 415m in length and 60 meters in width: a surface area of approximately 2.5ha.. It was required to address three issues: the resolution of an acoustic problem, the evacuation of rainfall deposits and the question of topography.

The sound issue caused by road traffic on the Ronda Hispanidad (Third Ring Road) affecting neighboring dwellings, requires the establishment of a sound barrier to include the whole north-western border of the park. The existing topographical ground level difference between ground-level of the new residential quarter and the ring road reaches a maximum height of 14m, where the containment of the terrain is resolved by means of a system of reinforced earth walls. This is made up of four steps apart from one another by 1.5m, composed of a galvanized steel mesh and large gravel stones, thus forming a sound barrier that will protect future residential developments in the area.

To the far south-west of the park, where no significant topographical difference is noticeable, the issue of sound containment is resolved by means of a Cyclopean wall 100m long with a maximum height of 10m. This wall is moreover conceived as an icon that characterizes the new neighborhood and also provides access to the underground square or mill basin situated in its extrados.

This laminar flow basin is designed to cope with the intense rainfall that affects the area, thus preventing floodwaters from emptying into the municipal network, whose diameter and capacity

are insufficient to deal with such heavy quantities of rainwater. This compound with its large surface area (3,150m²), whose use as a laminar flow space will be conditioned by the frequency and intensity of local rainfall, has been conceived and designed as an urban space or pedestrian square for most of the year and a welcome area of shelter from the unpleasant Cierzo wind which blows in this upper area of the city. Four stairs situated at the corners provide access to the underground square, connecting with the adjacent neighborhood and the city level. The incorporation of sufficiently wide ramps situated within the sound barrier wall gives access to service and maintenance vehicles and a more ample use of the compound.

Finally Venecia Park is a carefully planned topographical operation that complements the acoustic functions and flow-forming processes described above in addition to providing green spaces to the city. All these are structured spatially over the Ronda Hispanidad by means of staggered interconnecting platforms in a linear or extended link-up of little squares (hard and soft), viewing points protected with light metallic pergolas, extensive groves of pines and pedestrian ramps leading to the historic heritage site of Aragón's Imperial Canal. Héctor Fernández Elorza

项目名称：Venecia Park
地点：Sector 88/1, Pinares de Venecia, Zaragoza
建筑师：Héctor Fernández Elorza, Manuel Fernández Ramírez
合作者：Félix Royo Millán, José Antonio Alonso García, Antonio Gros Bañeres, Beatriz Navarro Pérez
甲方：Junta de Compensación del Sector 88/1
用地面积：25,000m²
竣工时间：2011.12
造价：EUR 2,598,799
摄影师：©Montse Zamorano (courtesy of the architect)(except as noted)

东南立面 south-east elevation

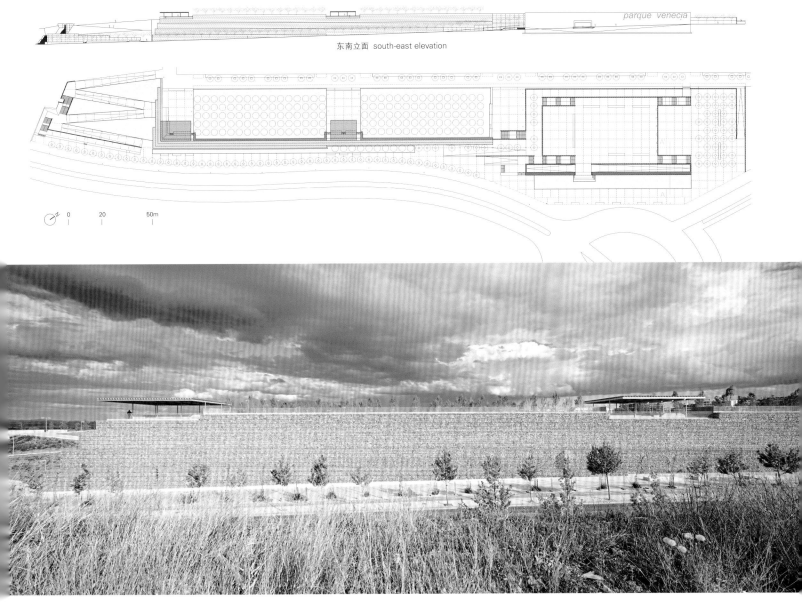

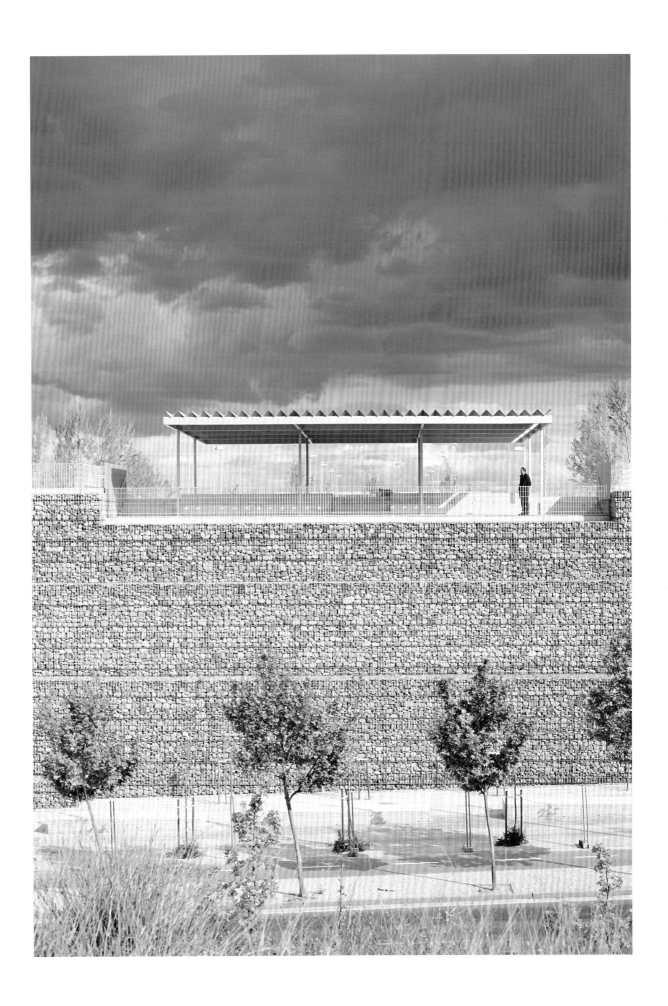

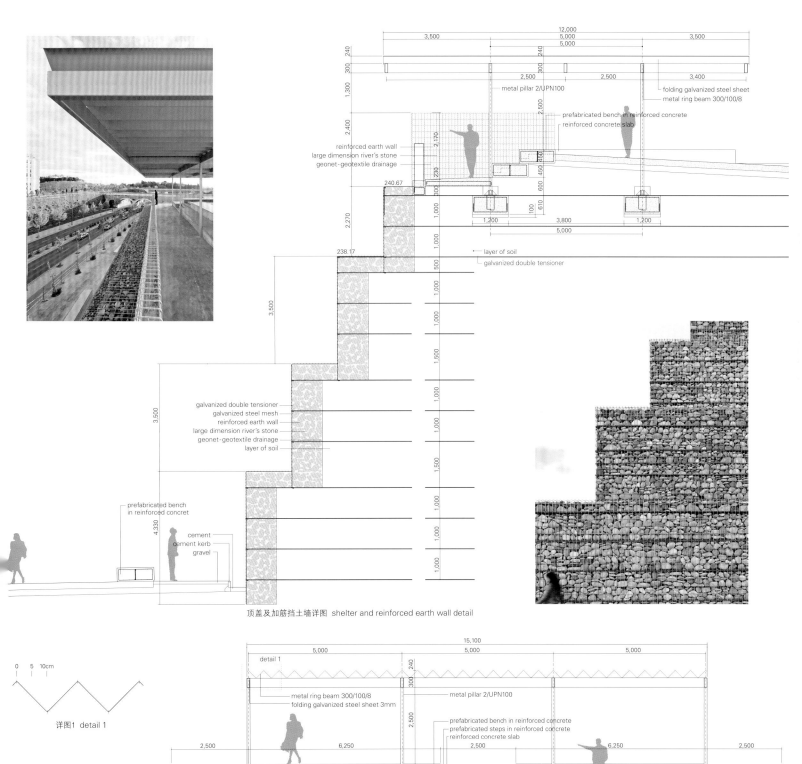

顶盖及加筋挡土墙详图 shelter and reinforced earth wall detail

详图1 detail 1

顶盖立面 shelter elevation

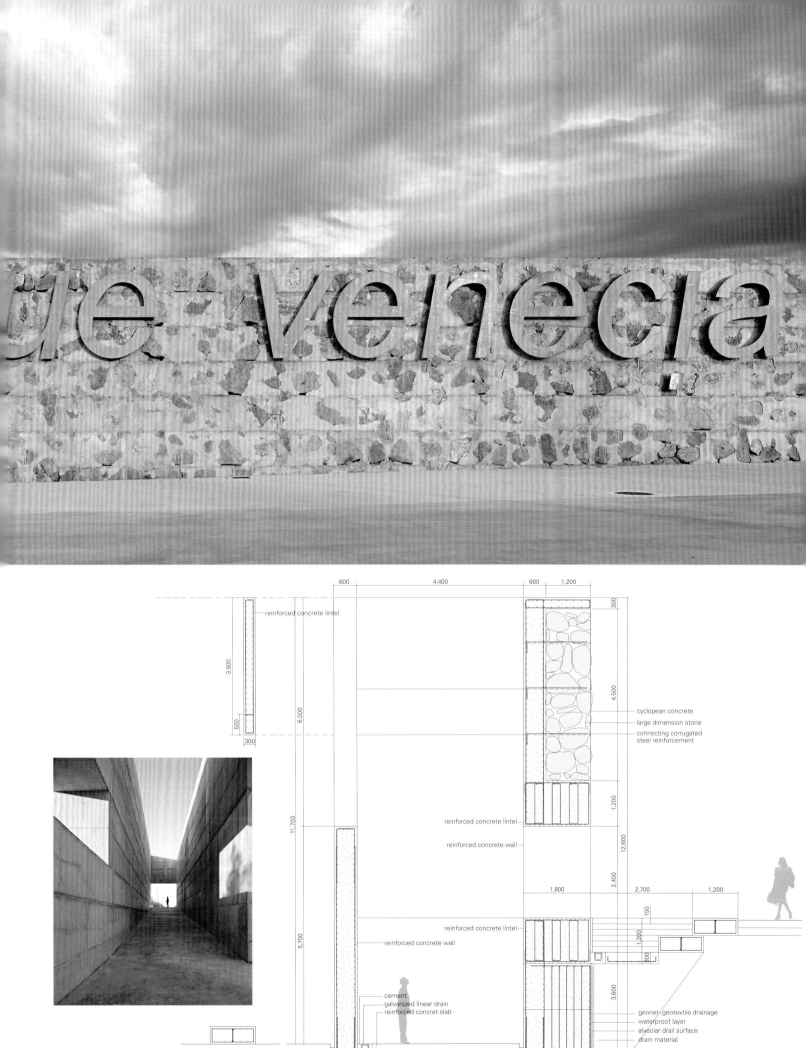

A-A' 剖面图 section A-A'

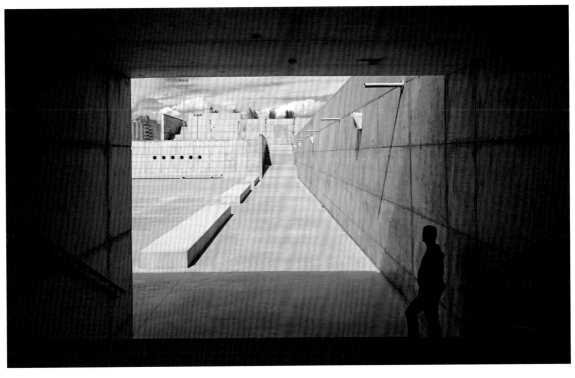

双子广场

北广场

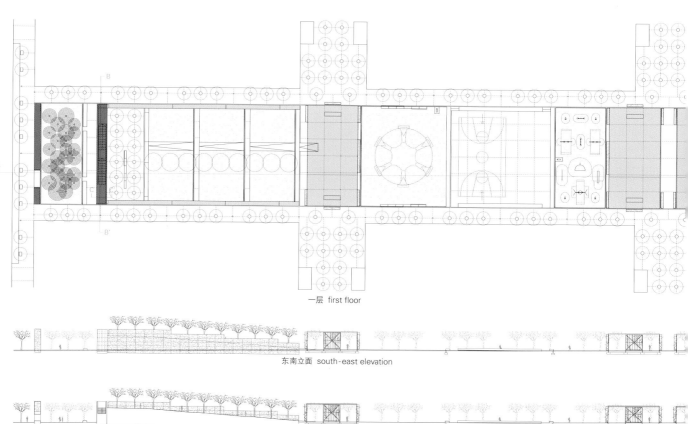

一层 first floor

东南立面 south-east elevation

A-A' 剖面图 section A-A'

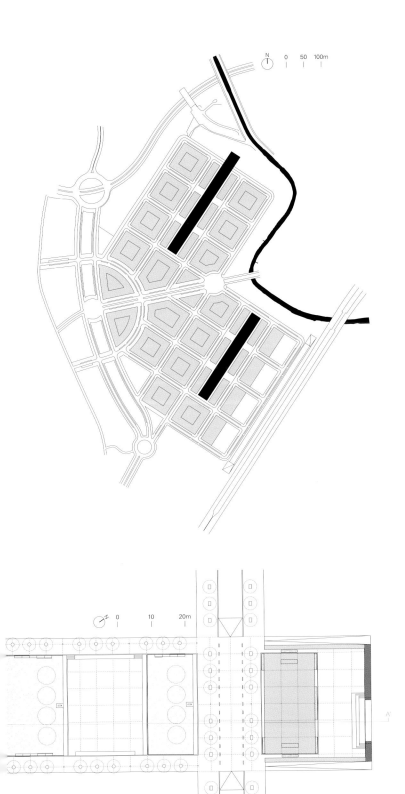

设计两个广场如同数字所暗示的那样有很多不便，设计规则也是成对的。同时设计任何两个事物都会不可避免地相互比较，找出相似和差异之处，对比优缺点，优缺点不仅是两个部分的特征，也是整体的特征。

萨拉戈萨88/1区的双子广场就是这样。广场需设计为相似的但同时也要有各自的特点的一对广场，换句话说，是双子广场但又不是完全相同。

这一区域的道路网对称排列，这两个广场打破了这一秩序，广场东北部地势较高，可以俯视阿拉贡皇家运河及周围景色，天气晴好时，可以目及远处的庇里牛斯山。两个广场一个在北，一个在南，广场其余三面由旁边新建的居民楼环绕。独具新特点的结构建立了新的序列，但并不完全相同。

由于新区内公共空间的形状特殊，相同元素处于不同位置上也决定了两个广场不会建成一模一样。

两个广场均宽40m，南广场长280m，占据了广场两边道路之间的有效空地，北广场横跨东北方的道路，一直延伸到皇家运河的绿化带，总长330m。南广场离皇家运河较远，比相邻的街道高5m；北广场与与南广场高度相同，朝向运河旁边的堤岸开敞，北广场北侧界墙是一堵巨石墙。南广场离运河较远，地势较高，可以远眺庇里牛斯山；北广场在边界墙上设计了一个巨大洞口，透过洞口可以看见附近的皇家运河。

这两个广场结合了各种元素：巨石墙、喷泉、遮光棚、高架平台、日光区、阶梯墙、绿色墙体、"硬地"广场、运动设施、儿童游乐场地。这些元素将南北两个广场组织为宽度相同但深度不同的带状结构。

因此，两个广场的东南面都有一堵长度相同、宽1.8m、高度不同的巨石墙，将广场与市区隔开。两个广场内各有两座喷泉，位置各异，因此它们之间的关系也不同。在这两个广场中，街道和广场高架平台地势之间的差异因修建了宽3m但高度不同的钢筋混凝土阶梯墙而得到了缓解，这些阶梯墙的内部还有一条意大利坡道。镀锌钢制成的遮光棚单独或成对设置，从广场的一面长墙延伸到另一面长墙，一直延伸到城市区域。除了沿着广场长向种植了树木，日光区的斜坡上铺设了草皮，遮光棚周围的金属笼及不锈钢钢缆内还种植了攀援植物。在竖直和倾斜面上利用植被的这种方式保证了最低的维护率和最大的视觉效果。

不同的带状活动区域通过长度与广场宽度相同的混凝土长椅分隔开。相同的长椅可以让在游乐场上玩耍的孩子、在树下乘凉老人或玩篮球的年轻人坐下来休息一会儿；白天的活动也是成对的。

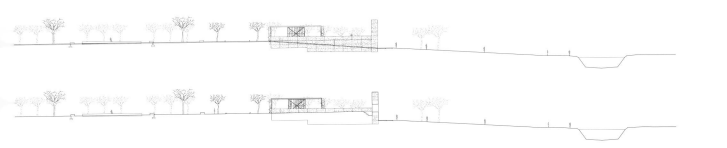

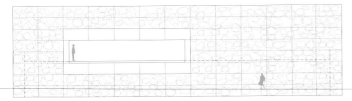
东北立面 north-east elevation

西南立面 south-west elevation

Twin Squares

Two squares: with all the inconvenience that the number implies and the rules of the game defined by a pair. Working with two of anything leads to comparison, defining similarities and differences; compared virtues and defects that not only define the two parts but also characterize the ensemble.

Such is the case of the two squares within sector 88/1, in Zaragoza. Squares constructed as a pairing, similar, but at the same time with their differences. In other words, twin squares, but not identical.

Within the symmetrical layout of the road network of the sector, the two squares break with the prevailing order with their north-eastern side overlooking the Imperial Canal of Aragón and the privileged views afforded by this upper part of the city and to the Pyrenees beyond on a clear day. One to the north and the other to the south of the sector, the two squares are surrounded on the remaining three sides by the buildings of the adjacent newly-built neighborhood. Constructions that will all bear the hallmark of the new established order, but will inevitably not all be the same.

Because of the configuration of these public spaces within the new district, the different positioning of the same elements that make up one and the other square predicates that they do not end up the same.

So while the two squares are equally wide, each measuring 40m, the southern square, 280m in length, fits the available space defined by the roads bordering two of its sides, while the northern square is set astride the northeastern road as far as the green belt of the Imperial Canal, a total of 330m in length. While the southern square is set back from the Canal Imperial and its ground level is 5 meters above the adjacent street, the northern square is at the same level and opens onto the embankment beside the canal where its northern flank is bounded by a Cyclopean wall. While the southern square is set back and raised to underline the views to the far-off Pyrenees, the northern square takes a step forward with a large window in its boundary wall to frame its views to the adjacent Imperial Canal.

It is the combination of the various elements: the Cyclopean walls, the fountains, the canopies, the elevated platforms, the solarium, the stair-walls, the green walls, the "hard" squares, the sports facilities and the children's playgrounds that organizes the distribution of the two squares into strips of varied depth and equal width.

Thus the south-eastern sides of both squares are closed off to the city by two Cyclopean walls of equal length and width 1.8m, but of differing height. Both squares hide a pair of fountains: double pairs differently positioned and therefore variable in their relationship. In the two squares the difference between street-level and the elevated platforms is resolved by various reinforced concrete stair-walls of varying height and 3m in width accommodating an Italian ramp in the shadow of the interior. Light canopies in galvanized steel are distributed singly or in pairs across the squares from one long side to another and continuing on into the urban district. Apart from the trees along the length of the squares, vegetation is limited to the sloping panes of the solariums and the vines whose growth is protected by the mesh cages and stainless steel cables surrounding the canopies. In this way maintenance levels are kept to a minimum but with maximum visual effect through the use of vegetation on both vertical and sloping planes.

The strips for distinct activities are separated from one another by concrete benches of the same length as the width of the square. The same benches provide seats for children beside their playground area, the old man under the shade of the tree or young people taking a rest from their game of basketball: daily activities in pairs. Héctor Fernández Elorza

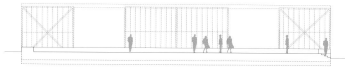

硬地广场立面 hard square elevation

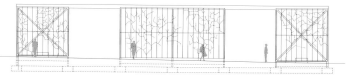

遮光棚立面 shelter elevation

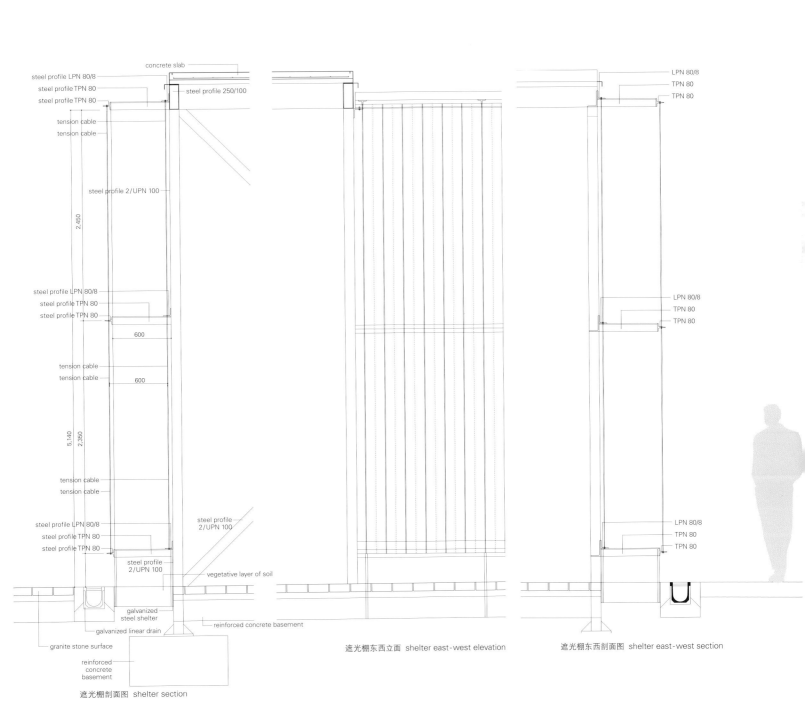

遮光棚剖面图 shelter section

遮光棚东西立面 shelter east-west elevation

遮光棚东西剖面图 shelter east-west section

157

项目名称：Twin Squares
地点：Sector 88/1, Pinares de Venecia, Zaragoza
建筑师：Héctor Fernández Elorza, Manuel Fernández Ramírez
合作者：Félix Royo Millán, José Antonio Alonso García, Antonio Gros Bañeres, Beatriz Navarro Pérez
甲方：Junta de Compensación del Sector 88/1
用地面积：50,000m²
造价：EUR 2,549,572
竣工时间：2011.3
摄影师：©Montse Zamorano (courtesy of the architect)

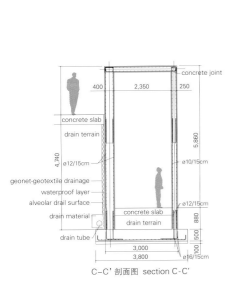

C-C' 剖面图 section C-C'

Valdefierro公园

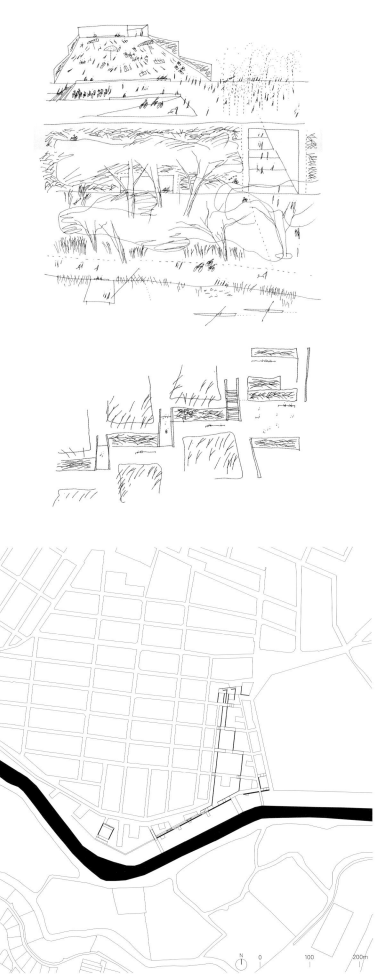

萨拉戈萨Valdefierro公园项目的主要设计特点是由其自身位置环境所决定的。

一方面，园址所在地的土壤降解严重。一块占地11公顷的L形条状土地北面和西面与Valdefierro区后方相接，南面是阿拉贡皇家运河，多年来，这里曾是一个砾石坑，后来成为垃圾填埋场，主要填埋城市的建筑垃圾废物。在如此大面积的区域内清理、运输、回收现有的残骸垃圾需持续的大量投资，将导致预期的工作量和资金失衡。

另一方面，地形环境非常特殊。从阿拉贡皇家运河到附近小区建筑物之间高度落差约为9m，这个差度使河床看起来比实际距离更远。

场地主要环境因素包括：砂砾层残骸（那时没人想到把这些巨大的砾石用作铺路石）、垃圾填埋场（主要是早期城市建筑工地的碎石废料）及场地自身的独特地形。这些因素使得建筑师采用一系列墙体构成的几何形式来构建这一项目。

砂砾与碎石用水泥混合，建成厚重的巨石墙。考虑到重力，这些无筋墙的厚度因高度不同而变化。这些墙把公园划分为阶梯，确定了公园的地形。原先地形的不利因素却成为了公园设计的一大亮点。

平面设计的其他部分也是通过几何方法来解决的。阿拉贡皇家运河附近和南端连接处只有一个梯度的高度差：巨石墙长210m，厚1.8m，高9m，连接了公园和河流。附近的居民区拥有一处视觉位置最佳的公共空间，凸显了大运河的美景。与此同时，巨石墙的深度使行人沿楼梯、坡道、长椅可以到达大运河，这些便利的设施安设在公园之内。与此相反，场地东面的空地使公园在高度上呈三个阶梯状。双排高4m、厚1.25m的巨石墙建筑结构顺应了三级梯形变化的几何状地势。

勾勒出地形的特色巨石墙是利用园址内的石头和土壤修建起来的。狭窄的钢筋混凝土墙将从公园到街区的横向行人通道（楼梯和坡道）与现有的街道网络相连，公路、公园、街区融为一体。墙的表层不同，功能也不同。钢筋混凝土墙细窄，巨石墙厚重敦实，二者相互补充，对比鲜明。模制构件的表面经过金属浇筑设备的处理后变得光滑闪亮，厚实的巨石墙表层粗糙，内部纹理由于旋转冠齿轮的磨损变得清晰可见。

简而言之，公园的地势呈阶梯状，新近栽种的树木很容易存活，可以抵挡凛冽的北风。水平状平台既方便居民到此活动，也满足了公园的布局需要。居民可以爬楼梯或走坡道，在不同高度的平台之间来回走动，楼梯和坡道均建在钢筋混凝土墙之间，沿着楼梯和坡道可以走到街区的街道。居民可以坐在雕在巨石墙上的长椅上休憩或者爬楼梯走坡道到达公园内部。这些土墙可以挡风，同时吸纳冬日的暖阳。树木在墙的映衬下显得郁郁葱葱，自然景观更美。墙上的洞口使人可以观赏公园的亮丽风景。墙壁弹回孩童的皮球，为老者遮阳。墙上爬满藤蔓植物，昆虫和小鸟将会在此安家，墙壁上也会布满各种涂鸦。这些魅力无限的手工艺术墙，就像巨大的地毯，由水泥和石头经纬线编织而成，成为交流的场所、讨论的空间、初吻的隐秘私地。

Valdefierro Park

The major decisions concerning the Valdefierro Park Project in Zaragoza were determined by the opportunities afforded by the context of the site itself.

On one hand, the soil where the park was to be situated was considerably degraded. An L-shaped strip of land covering 11ha., bordered to the north and west by the rear of the Valdefierro district and to the south by the Imperial Aragón Canal had been used for years as a gravel-pit and later as a land-fill site, mainly for waste from building works in the city. The clean-up, transfer and recycling of the existing debris in such a large area of the site would have required substantial investment, disproportionate to the vol-

ume and budget of the proposed work.

On the other hand, the topographical context is quite pronounced. Almost 9m of difference separated the height of the Imperial Aragón Canal from the level of the nearby buildings of the neighborhood; a difference that caused the riverbed to appear more distant than it really is.

Such determining contextual factors: the gravel-bed debris (with those large gravel stones which at the time nobody wanted to use as gravel), the land-fill site (composed mainly from the rubble of former construction works in the city) and the pronounced topography of the site, led architects to construct the project with the geometry of a system of walls.

The gravel and rubble were mixed with cement to construct very thick Cyclopean walls. These unreinforced walls, which on account of gravity vary in depth according to their height, distribute the layout of the site into terraces and determine the topography of the park. Thus the initial contextual problems are turned around

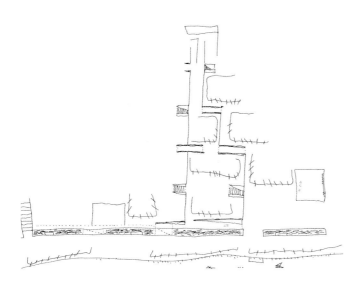

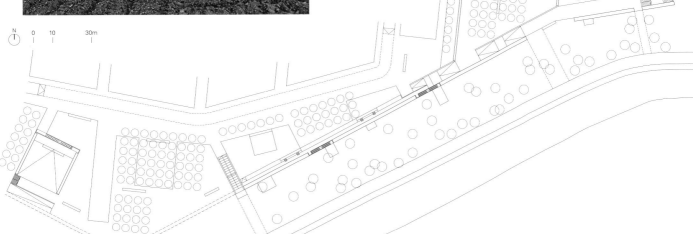

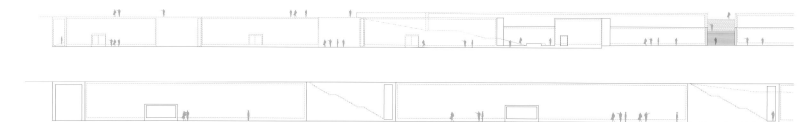

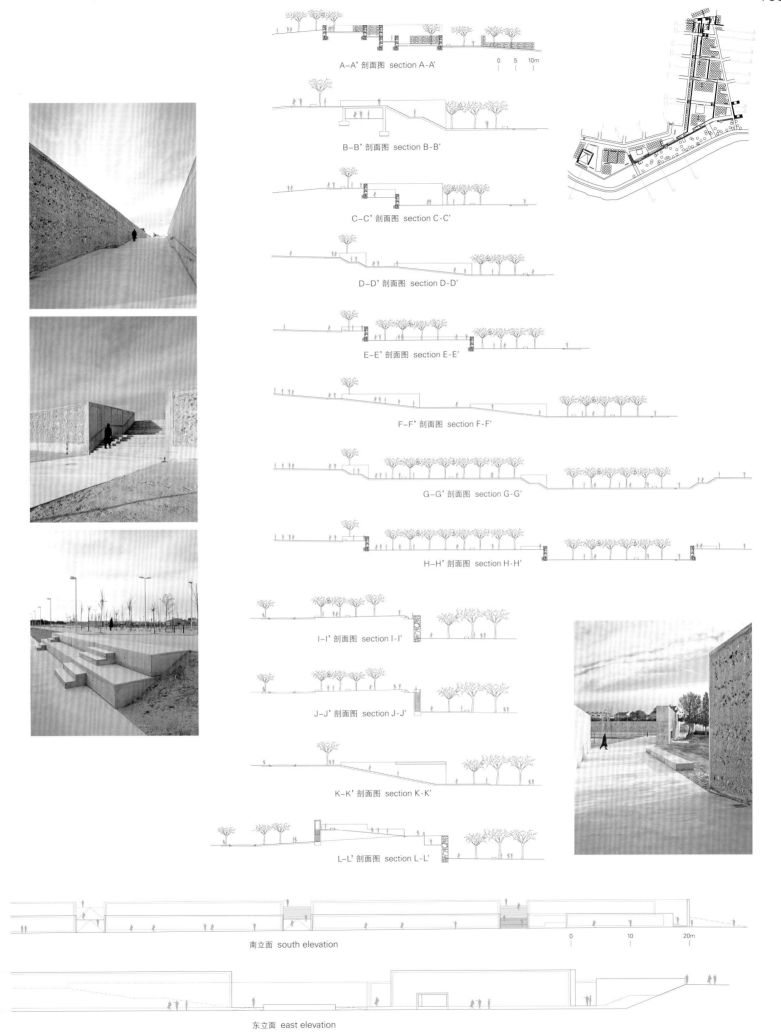

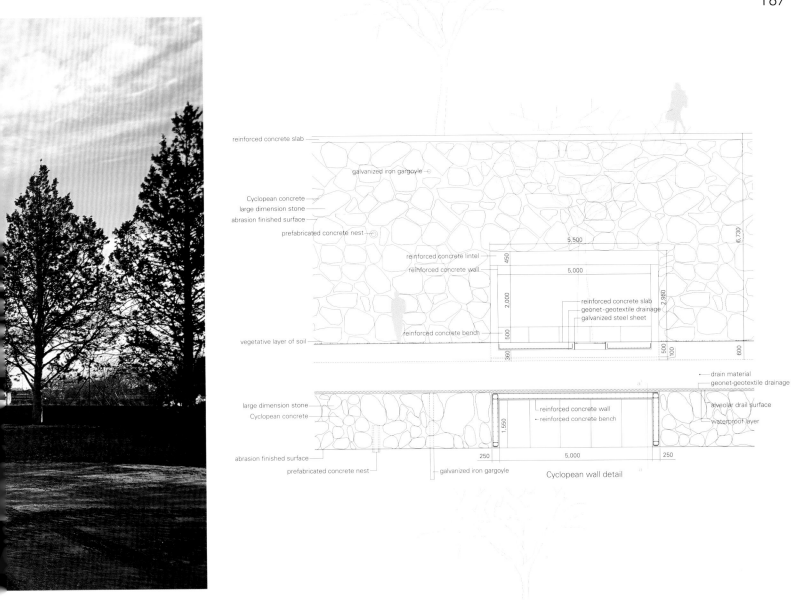

Cyclopean wall detail

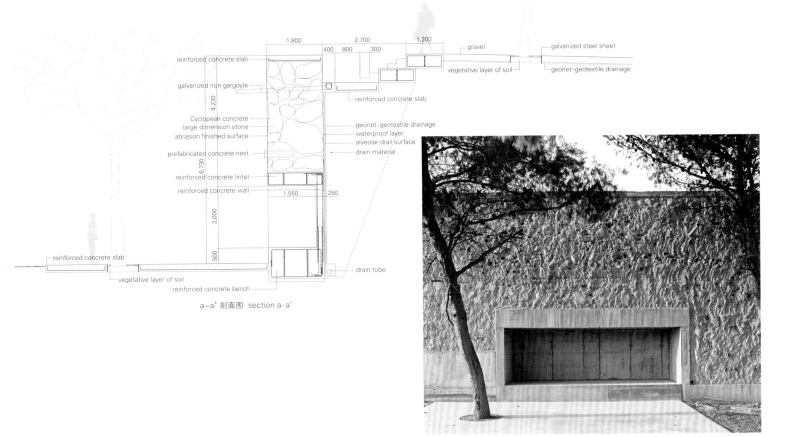

a-a' 剖面图 section a-a'

项目名称：Valdefierro Park
地点：Sector F-57/8, Barrio de Valdefierro, Zaragoza
建筑师：Héctor Fernández Florza, Manuel Fernández Ramírez
合作者：Félix Royo Millán, José Antonio Alonso García, Antonio Gros Bañeres,
建造商：Construcciones Mariano López Navarro, SAU
甲方：Sociedad Municipal zaragoza vivienda, SLU
表面面积：110,000m²
成本：EUR 5,010,000
施工时间：2009.9—2010.12
摄影师：©Montse Zamorano (courtesy of the architect) (except as noted)

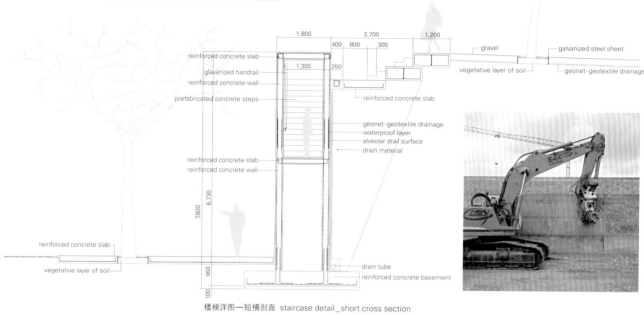

楼梯详图—短横剖面 staircase detail_short cross section

楼梯详图—长横剖面 staircase detail_long cross section

to favor the design itself.

The remaining layout is resolved geometrically. The proximity of the Imperial Canal and its link-up with the southern end is built with just one drop in level: a Cyclopean wall 210m long, 1.8m thick and 9m tall resolves the connection between park and river. The neighborhood thus benefits from a public space that is optimally placed and serves as a backdrop underlining the canal landscape, while at the same time the depth of the walls allows for the stairs, ramps and benches providing greater accessibility to the canal to be hidden within the construction. Conversely, the extent of space available on the eastern side of the site of the site allows for this area of the park to be distributed into three terraced levels; three terraces of variable geometry that adapt to the terrain by means of a double row of Cyclopean walls 1.25m in depth and 4m tall.

If the defining Cyclopean walls outlining the topography are constructed from the very stone and soil of the site, the transversal pedestrian connecting areas (ramps and stairways), from the park to the neighborhood, in continuity with the existing street network, is resolved by means of reinforced narrow concrete walls. Two different skins have very different functions. The slenderness of the reinforced concrete wall sections is both compensated by and in contrast with the chunky aspect of the Cyclopean masonry. The smooth, polished texture produced by the metallic casting of

the molded sections contrasts with the rough surface of the thick Cyclopean walls, whose internal texture has been revealed by the abrasive action of a rotary crown gear.

In short, the park has been built in terraces, on which the recently planted trees will easily grow, protected from the harsh north wind. The horizontal terraces are designed so that local residents will adapt their activities and needs to the layout of the park. These same residents will move about between the different levels using the stairways and ramps built between the reinforced concrete walls that continue into the streets of their neighborhood; they will sit on the benches carved into the Cyclopean masonry or they will make their way through the interior using the various stairwells and ramps. These earthen walls will provide protection from the wind and at the same time receive the welcome rays of winter sunshine; they will highlight the trees and surrounding nature, framing the landscape through their various openings; returning the ball to a child playing or providing shade to an elderly resident. Climbing plants will grow on its walls, insects and birds will make their nests here and graffiti will make its appearance. These infinite artisanal walls, like giant carpets, interweaving with the warp and weft of cement and stone, will serve as a backdrop for conversations, and discussions or as a secret hiding-place for first kisses. Héctor Fernández Elorza

细胞和遗传生物学学院

Héctor Fernández Elorza建筑师事务所 Héctor Fernández Elorza

阿尔卡拉大学细胞和遗传生物学学院的修缮和扩建工程受制于两个方面：一座既存的建筑必须保留下来，同时还要进行扩建，从而满足现代的需要和要求。

原有的建筑本是20世纪上半叶的第一座军用机场建筑，自20世纪60年代末现代的校园建立以来，学校克服了一些困难对其进行改造，使它符合大学的需求。在过去的四十年里，这座建筑断断续续地经历了一系列的小翻修。它需要进行修缮，而且它缺少细胞和遗传生物学学院所需的研究及教学的必要空间。

既存建筑旁有两条南北向的路，整个西面都是花园。要在保持原有建筑立面和结构的同时，将建筑表面积扩大为原来的两倍，从而满足如今的需要，这就意味着要在原有的三层建筑物上加盖一层，此外，还要在地块的西面新建一个悬挑部分。

依照老建筑的体积容量，当前建筑项目包括重组沿着东面的教学和研究实验室、公共服务区、楼梯，还有通往老建筑南面和北面的通道以及连接新楼和实验室西侧的主走廊。新修的悬挑部分面向西面，包括办公室、研讨室和会议室，这一部分和老楼分隔，通过天桥到达主通道，主通道为这一新建筑的中央区带来了光照和通风。

原有建筑的混凝土结构、柱子、大梁和砂浆用金属结构加固，在加固的大梁上方留出了安装技术装置及电气装置的分隔空间。顶层的扩建部分是金属结构的梁柱并覆盖着金属板，新的悬挑部分中有办公室、会议室及研讨室，都是由混凝土建造的。

建筑的电气及技术装置在实验室和办公室都可以看到，这些装置从宽阔的出入口平行排列至主交通走廊，连接到屋顶的机械装置。

原有建筑的北立面、南立面和东立面原封不动，只是修补了外墙的隙缝。不再使用的门窗用大块镀锌金属板封堵。南立面和东立面的门窗上装了打有深孔的8mm镀锌钢，均匀的光线可照进内部的实验室、洗手间和更衣室。

原有建筑的立面中，只有西立面进行了整修，改造成了聚碳酸酯立面，为所有的走廊提供照明。扩建的第四层立面的终饰是黑灰泥，墙上没有开墙洞，屋顶的天窗为顶层提供了足够的阳光。

建筑朝向西侧广阔的花园。混凝土悬挑部分内设有教职人员办

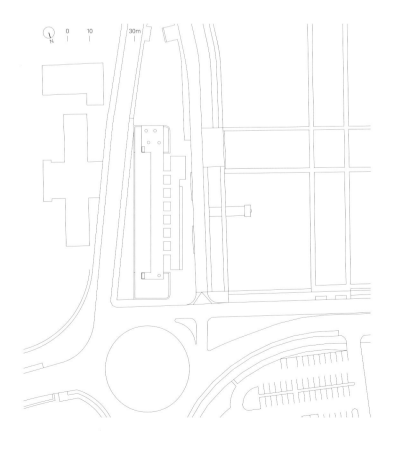

公室和会议室，建在四个墩座上，是厚梁结构，末端的悬臂可以俯视一层的公共服务区。这一层的辅导室和走廊由玻璃环绕，通过一条连续的混凝土人行道通往外面的花园，花园与室内外的关系不分等级。同样地，办公室和会议室面朝树林，避开西侧刺眼的眩光。办公室和会议室设有齐眼高度的墙洞，一面由落地百叶窗保护着，对面巨大的玻璃窗照进柔和均匀的光线，是一处幸运场地，在这里，人们可以通过看外面的树木来感知四季变换。

Faculty of Cellular and Genetic Biology

The project for the rehabilitation and extension of the Faculty of Cellular and Genetic Biology of the University of Alcalá was conditioned by two aspects: an existing building that had to be both maintained and extended in order to adapt to the needs and demands of today.

The existing building, originally the first aerodrome building for military use dating from the first half of the twentieth century, had been adapted with some difficulty from the late sixties onwards to suit the needs of the university with the founding of the present-day campus. Following a series of minor renovation-works over the last forty years, the building was in need of refurbishment and lacked the necessary space for the research and teaching required by the Faculty of Cellular and Genetic Biology.

There are two roadways running alongside the existing construction from north to south and extensive gardens on the entire western side. The need to maintain the facades and the structure of the original building and at the same time double its surface to adapt it to present-day requirements meant the addition of an upper floor above the existing three-story building as well as the addition of a new bay at the west of the plot.

The program of the current building in terms of the volume capacity of the old construction involved reorganizing the teaching and research laboratories along the eastern front, the common service areas, stairs and access ways to the north and south of the existing volume and the main connecting corridor with the new building to the west of the laboratories. The new bay, to the west, contains the offices, seminar- and meeting-rooms and is separated from the former building via connecting bridges to the main passageway, which in turn bring light and ventilation to this central zone of the new building.

The concrete structure of the original building, its pillars, beams

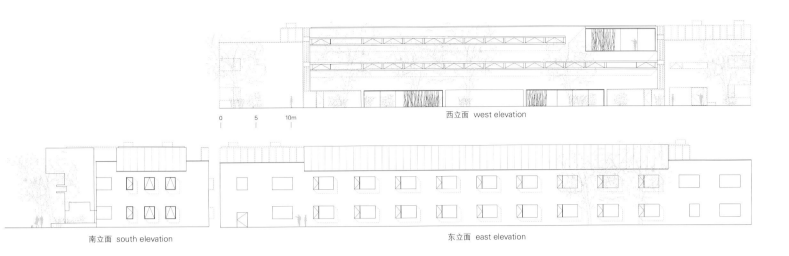

南立面 south elevation　　　　东立面 east elevation

项目名称：Faculty of Cellular and Genetic Biology
地点：University of Alcalá, Campus, Alcalá de Henares
建筑师：Héctor Fernández Elorza
合作者：Blanca Moreno, Carlos García Fernández, Irene Bodas
工程师：Raúl García Cuevas, Ignacio Delgado Conde,
Enrique Fernández Tapia,
施工：J. Quijano Construcciones S.L., Imaga Proyectos y
Construcciones S.A., Ferrovial S.L.
总建筑面积：3,540m²
造价：EUR 3,750,000
竣工时间：2012.6
摄影师：©Montse Zamorano (courtesy of the architect)

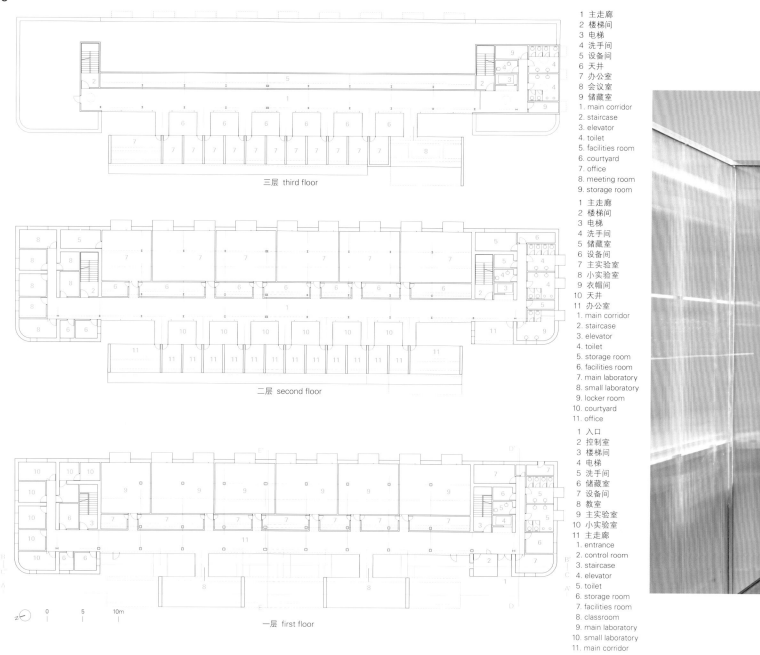

三层 third floor

1 主走廊
2 楼梯间
3 电梯
4 洗手间
5 设备间
6 天井
7 办公室
8 会议室
9 储藏室
1. main corridor
2. staircase
3. elevator
4. toilet
5. facilities room
6. courtyard
7. office
8. meeting room
9. storage room

二层 second floor

1 主走廊
2 楼梯间
3 电梯
4 洗手间
5 储藏室
6 设备间
7 主实验室
8 小实验室
9 衣帽间
10 天井
11 办公室
1. main corridor
2. staircase
3. elevator
4. toilet
5. storage room
6. facilities room
7. main laboratory
8. small laboratory
9. locker room
10. courtyard
11. office

一层 first floor

1 入口
2 控制室
3 楼梯间
4 电梯
5 洗手间
6 储藏室
7 设备间
8 教室
9 主实验室
10 小实验室
11 主走廊
1. entrance
2. control room
3. staircase
4. elevator
5. toilet
6. storage room
7. facilities room
8. classroom
9. main laboratory
10. small laboratory
11. main corridor

and screeds are all reinforced with a metal structure leaving the required separation zone above the reinforcement of the beams to allow for technical and electrical installations. The upper floor extension is built with a metal structure of beams and pillars and covered with sheet metal. The new bay housing the offices and rooms for meetings and seminars is built entirely in concrete.

The building's electrical and technical installations are all visible both in the laboratories and the offices and are organized from wide openings parallel to the main communications corridor and in connection with the machinery of the roof.

The north, south and east facades of the original construction remain intact except for the configuration of their apertures. The openings no longer required are closed up with large galvanized metal plates. The openings on the south and east facades are constructed with deep loopholes also in 8 mm galvanized steel, providing homogeneous light inside the laboratories, washrooms and changing rooms.

The only facade modified is that of the west of the original building, which is transformed into a polycarbonate facade thus uniformly illuminating the corridors. The facades of the fourth floor extension are covered with a finish of black plaster and deployed without any wall-openings thanks to the skylights illuminating this top floor from the roof.

The building looks onto the extensive gardens on its western side. The concrete bay housing the offices of the teaching staff and board room stands on four piers and a structure of thick girders with cantilevers at the ends permitting visual connection from the common areas on the ground floor. The ambiance of the tutorial rooms and corridors which are enclosed with glass at this level extends via a continuous concrete pavement outwards to the garden which is non-hierarchical in relation to interior and exterior. Similarly the offices and boardroom look out onto the trees and are protected from the uncomfortable glare of sunlight from the west. Fitted with eye-level openings and enjoying the protection of the deep brise-soleil of the floor above on one side and furnished with homogeneous and uniform light by the large windows of the opposite side the offices and boardroom have become a privileged spot from which one can observe the different seasons of the year reflected in the trees outside.

Héctor Fernández Elorza

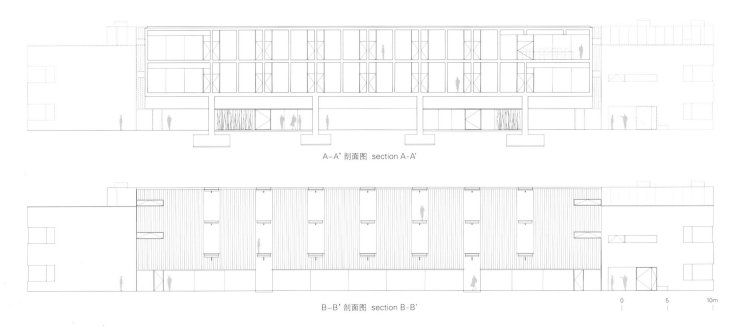

A-A' 剖面图 section A-A'

B-B' 剖面图 section B-B'

0　5　10m

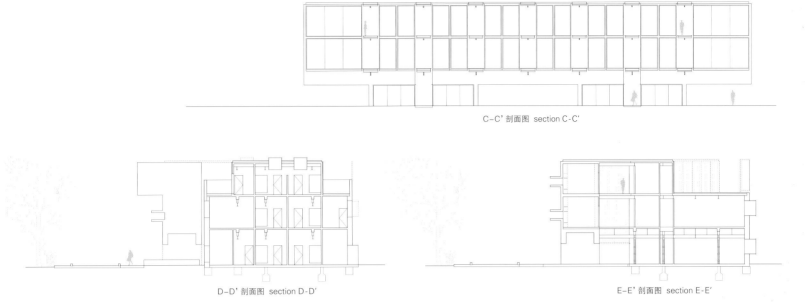

C-C' 剖面图　section C-C'

D-D' 剖面图　section D-D'

E-E' 剖面图　section E-E'

>>124

Ioannis Baltogiannis + Phoebe Giannisi + Zissis Kotionis + Katerina Kritou + Nikolaos Platsas

Ioannis Baltogiannis graduated from the school of architecture, University of Portsmouth with the title of master of architecture and has been working as practicing architect in Greece.
Phoebe Giannisi is an architect and poet, editor, and curator at the same time. Wrote six books and edited several books on architectural theory. Was a co-curator for the Greek Pavillion of the Venice Biennale 12th International Architecture Exhibition.
Zissis Kotionis is a professor of the University of Thessaly. Was also a co-curator for the Greek Pavillion of the Venice Biennale 12th International Architecture Exhibition together with Phoebe Giannisi.
Katerina Kritou has been practicing with Nikolaos Platsas based in Volos . Has been taking part in various architectural and art exhibitions.
Nikolaos Platsas is an architect and a Ph.D candidate of the University of Thessaly, school of architecture. Is currently involved in teaching undergraduate and postgraduate students at the same university.

>>58

Schwartz/Silver Architects

A founding partner of Schwartz/Silver Architects, Warren Schwartz received a bachelor of architecture from Cornell University and a master of architecture in Urban Design from Harvard University Graduate School of Design. Is the project designer of several of the firm's most honored projects, including the Shaw Center for the Arts and MIT's Rotch Architecture Library. Is a frequent juror for AIA awards and scholarships. Is actively involved in professional and civic organizations including the Boston 400, the Mayor's Task Force on the Future of the City. Is also a former design commissioner of the Boston Society of Architects.

>>38

Gonçalo Byrne Arquitectos

Gonçalo Byrne[right] is the founder and senior CEO of Gonçalo Byrne Arquitectos. For the last 35 years his works have been internationally recognized. Was a professor in Coimbra, Lausanne, Venice, Mendrisio, Leuven, and Havard.

Barbas Lopes Arquitectos

Patrícia Barbas[middle] and Diogo Lopes[left] have been working in partnership since 2003 and established Barbas Lopes Arquitectos together in 2006. Their realized and under development projects include public and private buildings and single-family housing. Also concentrated on exhibition and interior design. The practice is also engaged in collaborations with many architects including Peter Märkli and Gonçalo Byrne.

>>46

OS3 Arkitektura

Was founded in 2004 by Ainara Sagarna, Juan Pedro Otaduy and Maialen Sagarna. Ainara Sagarna Aranburu and Maialen graduated from School of Architecture in San Sebastian(SASS), University of the Basque Country(UPV-EHU) in 1999 and received Ph.D in 2010. Currently are professors at the same university. Juan Pedro Otaduy Zubizarreta graduated from the same university with the other two co-founders and also received a master's degree in 2011.

>>96

Schmidt Arquitectos

Is the architectural studio which has been operating by Martin Schmidt R.[left], Horacio Schmidt C.[middle], and Horacio Schmidt R.[right]. Until 2001, the office had been led by Horacio Schmidt C. for 30 years and before him by his father Horacio Schmidt G. and now has been passed to his sons. Specialized in the design and development of projects such as urban, corporate buildings, houses, interior, furniture and industrial design. Philosophy of Schmidt Arquitectos is based in the interaction of the ideas from the site and time where the projects is placed and active interaction with clients and their colleagues.

>>86

Cadaval & Solà-Morales

Was founded in New York in 2003 and moved to both Barcelona and Mexico City in 2005. Eduardo Cadaval holds a BA from the National University of Mexico(2000) and a master's degree from Harvard University(2003). Is an associate professor at ETSAB and a visiting professor at the University of Pennsylvania and the University of Calgary. Clara Solà-Morales holds a BA from ETSAB(2000) and a master's degree from Harvard University(2003). Is a head of graduate studies at the Barcelona Institute of Architecture, visiting professor at the University of Calgary and associate professor at Tarragona's School of Architecture.

>>106

acaa / Kazuhiko Kishimoto

Kazuhiko Kishimoto was born in Tottori, Japan in 1968. Graduated from Tokai University in 1991. Established Atelier Cinqu in 1998 and changed the name of the firm in 2006. Currently, he is lecturing at Tokai University and Tokyo Designer Gakuin College.

>>116

Schemata Architects

Was founded by Jo Nagasaka in 1988 and moved to Kamimeguro in 2007. Jo Nagasaka was born in Osaka and graduated from the department of architecture, faculty of fine arts of Tokyo University of the Arts. Designed not only buildings but many tables called "Flat Table".

>>66

Daniel Moreno Flores + José María Sáez

Daniel Moreno Flores graduated from the Pontifical Catholic University of Ecuador. Is studying a master's degree in Argentina in advanced architectural design. Was chosen as the winner of the award for the Best Young Work on the 6th Ibero American Architecture and Urbanism Biennale of Lisbon in 2008.

José María Sáez graduated from the Polytechnic University of Madrid. Is a specialist in architecture and architectural rehabilitation. Received the national award for architectural design for the 15th Pan-American Architecture Biennale of Quito in 2006. Is an active designer and invited teacher in several universities in Ecuador, Argentina, Peru, Spain and the United States at the same time.

>>26

3ndy Studio

Marco Mazzetto[right] and Alessandro Lazzari[left] established the 3ndy Studio in Venice after achieving their master's degree at the University Institute of Architecture in Venice(IUAV) in 1999. 3ndyStudio + Project was created in 2007 with the collaboration of Massimiliano Martignon[middle] who also graduated from IUAV in 1999. 3ndy Studio tries to characterize places which never had its own identity with plannings that work in their functional system and also with an aesthetic research.

>>132

Héctor Fernández Elorza

Was born in Zaragoza in 1972. Received a degree from ETSAM (Madrid Technical School of Architecture) in 1998 where he has been working as lecturing professor since 2001. His postgraduate studies were continued in Scandinavia, awarded grant by the Marghit and Folke Perzhon Foundation in 1999 and 2000. The Ph.D. studies are currently being finalized at the ETSAM with the doctoral thesis named "Asplund vs. Lewerentz". In 2000, he represented Spain at the Venice Biennale. Is an author and co-author of some books including *E. G. Asplund, Exposición de Estocolmo 1930*.

>>76
OAB – Office of Architecture in Barcelona
OAB – Office of Architecture in Barcelona was founded in 2006. One of the founders Carlos Ferrater[left] is a professor of architectural project design at the Polytechnic University of Catalonia. The other co-founder, Xavier Martí[right] graduated from the School of Architecture of Barcelona (ETSAB) in 1995.

ADI Arquitectura
ADI Arquitectura is the result of collaboration of the two architects Carlos Escura[left] and Carlos Martín[right] since 1978. The firm was formed in 1995 and has been completed over 800 projects. Has won numerous competitions and received many awards for the past several years.

Nelson Mota
Graduated from the University of Coimbra in 1998 and received a master's degree in 2006 where he lectured from 2004 to 2009. Was awarded the Távora Prize in 2006 and wrote the book called *A Arquitectura do Quotidiano, 2010*. Is currently a researcher and guest lecturer at the TU Delft, in the Natherlands. Is a member of the editorial board of the academic journal Footprint and also one of the partners of Comoco Architects.

Jesús Donaire
Jesús Donaire graduated from the ETSAM in 2000 and Columbia University in 2008 where he obtained a master's degree in advanced architecture design. Has designed and built a single house in Cáceres, several exhibition designs and has been finalist in two national competitions. Has been collaborated with architects Jesús Aparicio and Alberto Campo in Madrid and David Chipperfield in London. His theoretical investigations and built projects have been published nationally and internationally. Has written articles for national and international magazines and has interviewed architects around the world such as Andrés Jaque and Toyo Ito.

C3,Issue 2013.2
All Rights Reserved. Authorized translation from the Korean-English language edition published by C3 Publishing Co., Seoul.

© 2013大连理工大学出版社
著作权合同登记06-2013年第45号

版权所有·侵权必究

图书在版编目(CIP)数据

锚固与飞翔：挑出的住居：汉英对照 / 韩国C3出版公社编；于风军等译. —大连：大连理工大学出版社，2013.4
ISBN 978-7-5611-7759-4

Ⅰ.①锚… Ⅱ.①韩…②于… Ⅲ.①建筑设计－建筑理论－汉、英 Ⅳ.①TU201.1

中国版本图书馆CIP数据核字(2013)第064521号

出版发行：大连理工大学出版社
　　　　　（地址：大连市软件园路80号　邮编：116023）
印　　刷：精一印刷（深圳）有限公司
幅面尺寸：225mm×300mm
印　　张：11.75
出版时间：2013年4月第1版
印刷时间：2013年4月第1次印刷
出 版 人：金英伟
统　　筹：房　磊
责任编辑：张昕焱
封面设计：王志峰
责任校对：张媛媛

书　　号：ISBN 978-7-5611-7759-4
定　　价：228.00元

发　行：0411-84708842
传　真：0411-84701466
E-mail: 12282980@qq.com
URL: http://www.dutp.cn